*Did God Kill the King of Sodom?*

# Did God Kill
# The King of Sodom?

## By
## Walt Cross

Dire Wolf Books
Stillwater, Oklahoma

Dire Wolf Books
502 East Liberty Avenue
Stillwater, Oklahoma 74075-2630

# First Edition

For information about permission to reproduce sections from this book, write to Permissions, Dire Wolf Books 502 E. Liberty Avenue Stillwater, OK 74075-2630

Cataloging Data

Cross, Walt 1949 –
*Did God Kill the King of Sodom?*: Walt Cross – 1st ed. ISBN: 978-0-9850996-2-6
1. City of Sodom 2. Cities of the Plain 3. Kikkar area of the Levant 3. King Bera 4. Ancient Near East history 5. Asteroid impacts – Nonfiction. I. Title

Dust jacket and book design by Walt Cross.
MANUFACTURED IN THE UNITED STATES

*For my son Justin because of his love for ancient history and its many attendant mysteries.*

*"Oh my son Absalom, my son, my son Absalom! Would that I had died for thee, O Absalom, my son, my son!"*

*The lament of King David*

Acquiring knowledge and applying it is what allows us to separate myth from reality.

-Walt Cross

# Introduction

Why write this book now, before the excavation of Sodom is completed? Outside of the *immediately* important historical implications of finding the site of biblical Sodom, and discovering its manner and date of destruction, there are other reasons.

The location of Tall el-Hammam, the ruins of what was once a large and powerful city-state of the Jordanic civilization, lies east of the Jordan River and is therefore in the Hashemite Kingdom of Jordan. The volatility of the entire Near East and in particular the whole of the Levant is well known. At almost any given moment the excavation could be halted for political or military reasons. Any cessation of the archeological process at the tall could last for years or even decades. So this is why I want to present before the reading public, what I see as an ongoing and very real threat for all cities of the Earth. This threat destroyed the city of Sodom four thousand years ago, and it came close to destroying a major modern city just last year.

For these listed reasons, and the fact I enjoy telling an interesting history story, I chose to write this book, at this time.

# Prologue

In 1849 British archeologist Austen Henry Layard discovered the clay tablet library of King Ashurbanipal, the last king of the Assyrian Empire. Ashurbanipal ruled in his capital city of Nineveh, near the present day city of Mosul in Iraq, from 668 to 627 BCE or 2686 years ago.

Among the thirty thousand plus tablets is the one depicted on the cover of this book and housed in the British Museum. This artifact, according to engineers Alan Bond and Mark Hempsell is an astrolabe copied from an even more ancient artifact. I suggest this artifact was in fact copied from one created during the reign of King Hammurabi nearly four thousand years ago.

This planisphere is a star chart analog computing instrument that depicts, in precise detail, the direction of travel and angle of descent of the passing and eventual impact of an asteroid. This suggests that the asteroid's passing and impact took place and was observed by astronomers during Hammurabi's time as king, the very time that the Cities of the Plain were destroyed.

Although approximately forty percent of this astrolabe, which is usually comprised of two disks, is missing or damaged, I believe this clay tablet, written in cuneiform, is nothing less than a contemporary record of the Biblical destruction of Sodom and Gomorrah. If this assumption is correct, this written record combined with the other facts outlined in this book, brings this event out of the realm of myth and into the light of science.

## Acknowledgements

Without the knowledge, insight, and hard work of Dr. Steven Collins, and his many associates including scientists, historians, students, volunteers, and workers, this book would not exist, or have a reason to exist.

These exceptional individuals, supported and hosted by the Trinity Southwest University Department of Archeology and the Department of Antiquities of the Hashemite Kingdom of Jordan, are deserving of our support. If you can, I urge you to support the Tall el-Hammam Excavation Project.

Additionally, I acknowledge the assistance of NASA, The Canadian Space Program, the European Space Agency, the B612 Foundation, and many others who are keeping an eye on the sky for all of us.

My one wish would be for the scientists of the excavation project and the scientists of the various space agencies and programs to get together and discuss the destruction of Sodom and the other Cities of the Plain. For what happened in the Kikkar valley four thousand years ago has deep and abiding implications for both our future and for today.

# CONTENTS

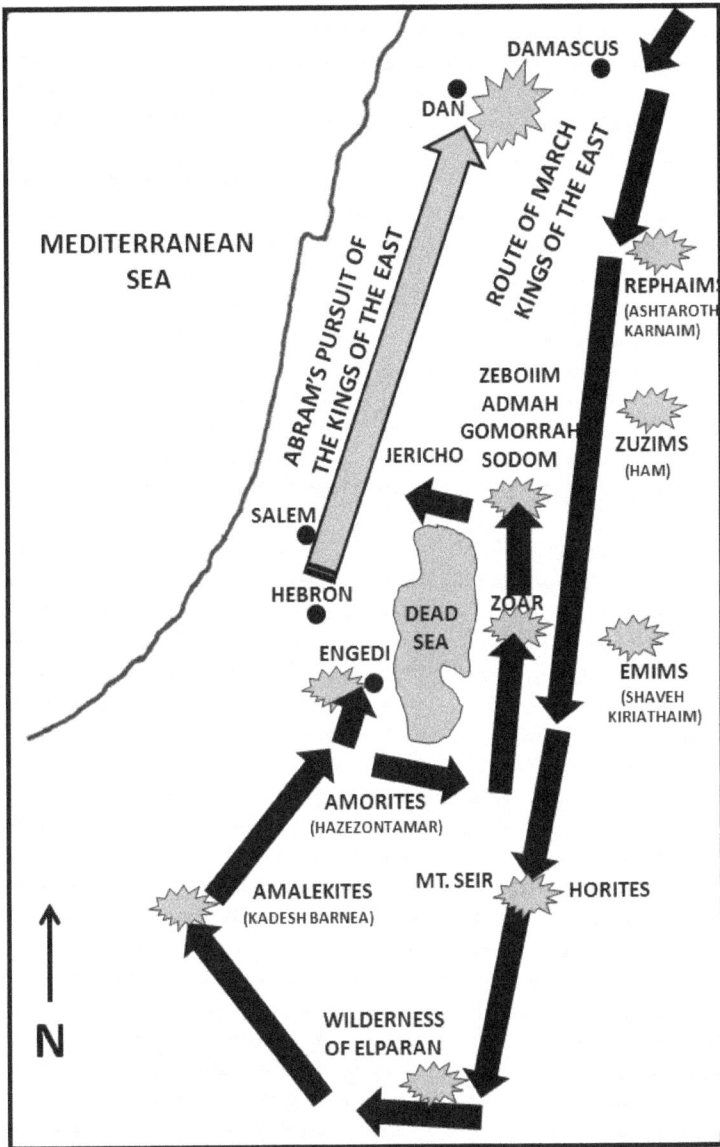

The route of march of the Eastern Kings in their campaign against the Land of Canaan and the Transjordan Plateau.

# Part I
# The Jordanic Civilization

# Chapter I
## The River Jordan

The Jordan River torturously makes its way from the fresh waters of Lake Galilee south to the saline brine of the inland Dead Sea. Nelson Glueck, a twentieth century archeologist describes the river's fate as

*...inevitably to perish in the 'Sea of Lot' as the Dead Sea is called by the Arabs.*

The river Jordan lives up to its name which means, 'river that goes down'.

The river flows through the Great Rift of the Levant, the same rift that forms the Nile River in pharaoh's far off Land of Egypt. The land of the Jordan is a land of earthquakes, the destroyer of cities as large as Jericho, of landslides that dam the flow of the river causing it to leave its banks to find a way around the boulder strewn path. Yet over the eons the earthquakes are a tool of the creator as well as that of the destroyer, forming the rift and its streams that bring life giving waters to its valleys and allowing the creation of civilizations.

The Jordan flows swiftly on its journey south from Galilee. Rapids, cascades, and whirlpools appear periodically along its two hundred mile length that only covers an actual land distance of sixty-five miles. The river seems to tarry as long as it can on its journey to oblivion in the Dead Sea. Its looping coils bend back and around forming ecological niches of jungle and well-watered open spaces. Its bounty has much to offer to the parched land through which it flows as it provides vital moisture to nourish its own civilizations.

Glueck notes that the general assumption of researchers has been that the Jordan River Valley has never been populous, yet recent discoveries, relatively speaking, question such assumptions. In Glueck's words:

*"Our recent explorations in the Jordan Valley have shown beyond doubt that it was once densely inhabited. Large and small settlements dotted the land. Excellent pottery was produced. Highly intensive agriculture was practiced. Thriving civilizations flourished. On the east side of the Jordan alone we discovered more than seventy ancient sites, many of them founded more than five thousand years ago, and some of them earlier...*

Glueck was so close to the remains of Sodom and Gomorrah that he likely trod upon and searched the surface of the ruins of those very famous cities. He goes on to add that the Jordan Valley was not only one of the first settled sections of the country, but that it was also one of the richest.

4

These different ruin sites, many of them the remains of the Cities of the Plain, are now barren reminders of extinguished life, now useful only as places to bury the dead. But the neglected and virtually forgotten mounds of debris piled into manmade hills were once houses comprising neighborhoods alive with the sounds of living people. Men and women once gathered inside their boundaries to discuss the day's events amid the hubbub of traffic and commerce and daily worship of long forgotten gods. Today, clues still lay scattered about in the form of pottery sherds, from once beautiful and elegantly constructed bowls, jars, and jugs now the wasted labor of a time past, ignored but by curious archeologists and tourists.

Combined with the mighty Jordan, streams flowed then and still today into the valley adding to the bounty of water that contributed to some of the very earliest settlements of man. Among these settlements and cities was numbered what the *Bible* calls the Cities of the Plain. And among these cities were one of the most well known combination place names, the cities of Sodom and Gomorrah.

Glueck expounds generously on the City of Jericho:

*The story of civilization might well start with the words 'And in the beginning there was Jericho'... Groves of fat trees flourished, and a heady wine was made of grapes ripening in the sub-tropical heat... Nothing has ever grown in the weird crisscross of chalky hills, which form a cruel belt of no man's land between the carefully planted acres of the Plain of Jericho and the wild Jungle of the Jordan below it... The goddess of the moon, to whom it was early dedicated, blessed it with unrestrained bounty... raiders from Arabia [the Transjordan] ...filled their bellies with the grain they had neither sown nor reaped, eaten the fruit of trees they had not tended, and carried off what they could not consume...*

Only a few short miles below Jericho and after flowing through its delta, the Jordan ends its life in the bitter waters of the Dead Sea.

# Chapter II
## Sodom's Mother City

### Tulayat al-Ghassul

Some researchers, including Dr. Steven Collins, author of *Discovering the City of Sodom,* think it a good probability that Tulayat al-Ghassul is the Copper Age (Chalcolithic) precursor city of *Biblical* Sodom. It is thought that some naturally occurring phenomenon, perhaps dealing with the loss of a water source, may have forced the inhabitants to move their city. The logical thought process then is that the inhabitants of Tulayat al-Ghassul (the modern name of the ruin) then moved north and founded Sodom. Assuming that the inhabitants are the genetic and cultural forebears of the people of Sodom, looking at Tulayat al-Ghassul should give us insights about them.

Tulayat al-Ghassul was a large city with its houses made of mud bricks, wood, and reeds. Some stone foundations are extant in the ruins as are well-defined courtyards. There is little doubt that Tulayat al-Ghassul was the dominant city in the Southern Levant, as its daughter city Sodom, would be in the future. In fact, it appears Sodom just assumed the industry and trade when the population left Tulayat al-Ghassul. It is likely it also assumed the same name as its parent metropolis.

7

This earlier Sodom developed trade with Egypt and Mesopotamia which included an area that contained parts of modern-day Syria, Turkey, Iraq Iran and the rest of the Land of Canaan. Items traded included copper and later bronze tools smelted from an alloy of tin and copper.

The waters of the Jordan River and other sources allowed intensive agriculture along the river banks. This included grains such as barley, wheat, legumes, olive trees, and grape vineyards. Pastoral pursuits practiced by nearby semi-nomadic communities supplied animal protein to the city dwellers. This likely included goat, perhaps sheep, and some meat from cattle. There may also have been a market for what is today referred to as bush meat or wild game.

These semi-nomadic groups eventually settled into permanent villages and hamlets, found themselves subjects of the local king, and came to embrace the local gods as well.

It appears to me that the people of Tulayat al-Ghassul were perhaps the original historical people and may have evolved from the Natufian culture (13,000 to 9,800 BCE). It therefore follows quite logically that the people of Sodom, Gomorrah, Admah, Zeboiim, Jericho and possibly Bela were the descendants of this original culture. If this is so, then the impact of the destruction of the Jordanic Civilization was an even greater tragedy for early civilization than is recognized today. It cries out for more investigation and understanding.

The loss of Sodom and the other Cities of the Plain may mark the end of a very unique culture stemming from the earliest of times.

What was striking to the archeologists that excavated the Tulayat al-Ghassul site were the paintings of masked men, stars, and geometric designs found on many interior house walls. The immediate thought was that these were religious paintings.

Interior wall painting from Tulaylat al-Ghassul found during the 1929 to 1934 excavations. Some have called this image a star, but my first impression upon viewing it was that it's a depiction of the Sun. These paintings were lost during an ill advised attempt to remove them from the wall. Other symbols surround the star including what looks like a mask. One theory is that the interior star is some kind of solar calendar.

Rivka Gonen wrote regarding these paintings[1]

*"Fresco fragments were executed in many rooms and it was assumed that the paintings were executed in a domestic context. These wall paintings, it should be noted, were not single creations. Rather, they were continuously plastered over and repainted... It has recently been speculated that the painted rooms were in fact shrines situated within the dense building clusters...The most remarkable segment is the star fresco. It consists of a large eight pointed star, 1.84 meters in diameter. Its center is composed of concentric circles and contains two more eight pointed stars. The innermost star is white on a black background, the second is white, bordered with black, on a red cross-hatched background; and the outermost star has alternating red and black rays. The execution is most precise."*

---

[1] The Archaeology of Ancient Israel 1992, Chapter 3 *The Chalcolithic Period.*

Face mask painting also found at Tulaylat Ghassul. It is possibly religious in character.

It appears that cultural rituals during the existence of Tulayat al-Ghassul developed independent from either Egypt or Mesopotamia. Again, it follows that such cultural rituals in the

This Tulayat al-Ghassul wall painting is thought to be a procession of three religious men walking toward a temple.

city of Sodom were likely unique from these two cultural dynamos as well. The question that begs to be asked is, if Sodom and its allied cities had survived, would they have emerged as a real counter balance to both the Egyptian and Mesopotamian civilizations?

A 2004 article[2] reported the results of radiocarbon dating of materials from Teleilat al-

Ghassul. The article explains that the testing indicates an end of habitation date of 3,900 to 4,000 BCE (Before the Common Era) for Tulayat al-Ghassul. It follows then, that Sodom was founded at or near that very same date.

The test materials were gathered from four separate areas of the ruins and consisted of both cereal grains and olive stones (pits).

In retrospect it seems doubtful that the entire population would uproot and move at the same time. Likely, families and groups of families moved as new mud brick houses were completed at the site of the new city. Soldiers would have had to provide security as the walls of the city were built. It also seems plausible that the lower city was built first with the upper city added later, perhaps built upon earlier construction. There is evidence that some townspeople left in haste. But it does not appear to be a widespread phenomenon. Perhaps as the end of habitation neared there was a hasty retreat for some reason such as interlopers (perhaps the ever present Bedouin) moving in. The continuing excavation of Tall el-Hammam should provide a definitive answer to some of these suppositions.

---

[2] *Proceedings of the 18th International Radiocarbon Conference* Vol. 46 Nr 1, 2004 pp 315 – 323.

Tulayat al-Ghassul was so striking to the archeologists excavating it that the culture and the age of mid-Chalcolithic Levant are referred to as The Ghassulian. I think it likely that Tulayat al-Ghassul dominated the Kikkar plain and laid the foundations for Sodom to continue that dominance for centuries to come.

One of the inferences that archeologist Dame Kathleen Kenyon drew about Tulayat al-Ghassul was that it was an intrusive culture not indigenous to the Levant. Her conclusion was that the people of Tulaylat al-Ghassul came to the Levant from either the east or the northeast. Of course at the time she had no knowledge of the close by location of the ruins of Sodom. However, her thoughts do mirror my own regarding the uniqueness of the Jordanic civilization's culture. It seems clear to me that Sodom and the other Cities of the Plain were neither wholly Canaanite nor Amorite. A view that I expect future excavations will demonstrate. Kenyon does recognize however that there is a direct succession of the Ghassulian culture to the time of the full Bronze Age.[3]

---

[3] *Archeology in the Holy Land* 1960 pp. 81 – 83.

# Chapter III
## Bronze Age Levant

Despite a number of different tribal names, the two major groups of people during the Bronze Age of the Levant[4] were Canaanites and Amorites. It is generally accepted both these groups are Semitic in origin. Some of their tribal names may sound familiar from passages of the *Bible*. Among these names are Hittites, Gibeonites, Hivites and Jebusites as well as many others. With some noted exceptions it seems correct to state that Canaanites tended to settle on the plains while the Amorites preferred the highland areas.

That portion of the Levant termed the Land of Canaan stretched from the Mediterranean coast to the Jordan River. However the people tended to be more dispersed than that, with some Amorites west of the Jordan and some Canaanites east of it. But in general the Amorites lives in the hills east of the river with Canaanites occupying the Jordan River valley later referred to as the Kikkar Plain.

---

[4] I use the term Levant because it is mostly a geographic reference whereas Palestine invokes a political as well as a geographic reference.

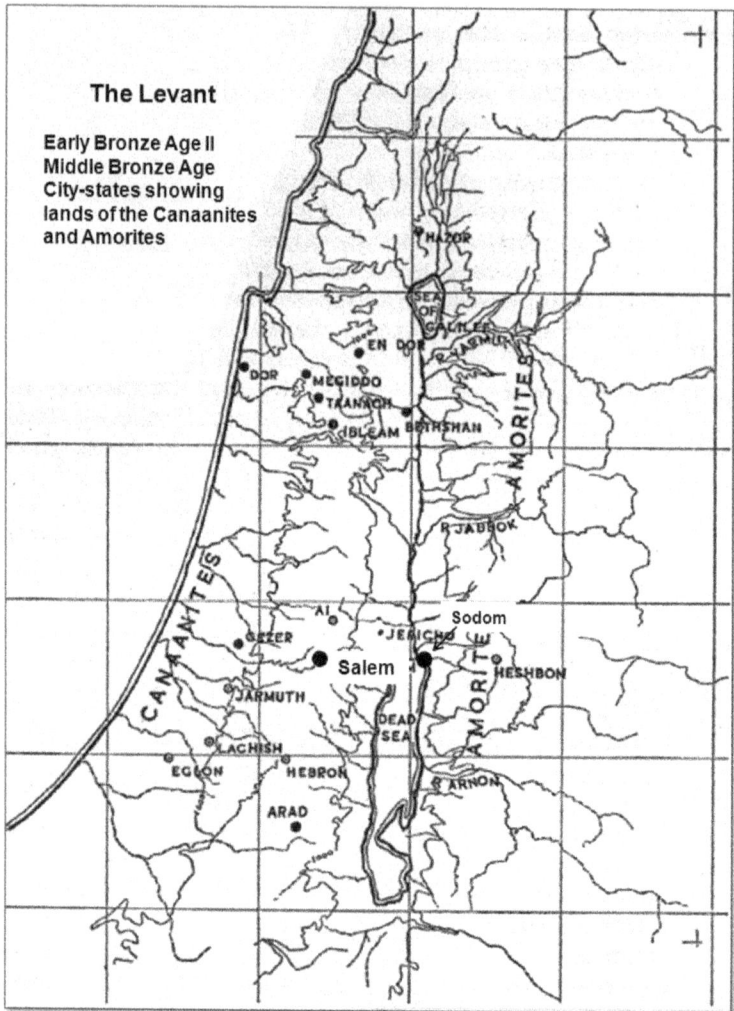

This map shows the general location of the two major groups of people in Bronze Age Levant, the Canaanites and the Amorites. Adapted from *Amorites and Canaanites.*

Biblically the Canaanites are described as the descendants of Canaan, a grandson of Noah. The

Phoenicians, who gave us the phonetic alphabet, were known as Kinahna which roughly seems to translate to Canaanite.

The Amorites referred to in Akkadian as the Amurru and in Egyptian as the Amar, are thought to have originated from the mountains of eastern Syria. Their immigration east and south through Mesopotamia and into the Levant was likely triggered by a major drought. One of their more famous cities was Babylon, destined to be ruled by King Hammurabi. This king is later referred to in the *Old Testament* as Amraphel, King of Shinar of whom we will hear more of.

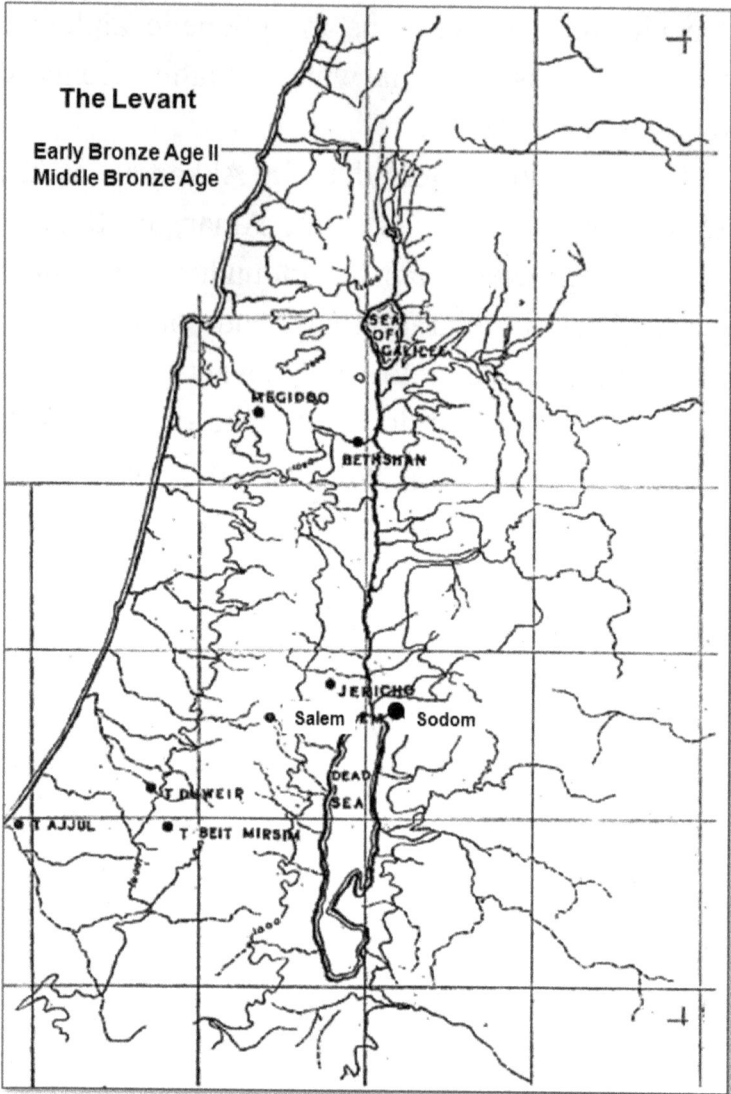

This map shows the location of some important cities of Bronze Age Levant. Adapted from *Amorites and Canaanites.*

The Amorites are described as a tall and powerful people and some references claim they were fair haired. Despite this description the Amorites were of Semitic origin. We will hear more of them later.

The Canaanites and the Amorites began the urbanization of the Levant. For the first time walled towns appeared over the breadth of the land, one of the earliest of which was Jericho, located just west of the Jordan River and in the Kikkar Plain. These towns for the most part were similar in their planning, establishment of religious centers, and in their production of pottery. Raiders who came primarily from the north and northwest with migrating groups of people prompted the building of town defenses that included walls, ditches, and eventually steep inclines. These groups of wandering people were collectively referred to as Habiru, a term thought to have evolved into the modern word Hebrew. These people would have a profound impact on the Levant.

Even mighty Egypt began to feel the pressure of these migrating peoples who banded together and pushed their way into the Nile Delta. The Egyptians called them the Hyksos, and along with their superior weapons and chariots, they brought their Storm God whom they called *Yahweh.*

A rendition of the destruction of the Amorite army by artist Gustave Dore. The Biblical account is that more Amorites were killed by *Yahweh* casting down stones from heaven than by the Habiru warriors.

Weapons fashioned from copper or bronze are scarce in the Early Bronze Age, but by the Middle Bronze Age they have become somewhat common with daggers the most common of all. So many males were buried with their daggers placed in either a waist or shoulder scabbard/sling that this type of burial became known as a dagger tomb.

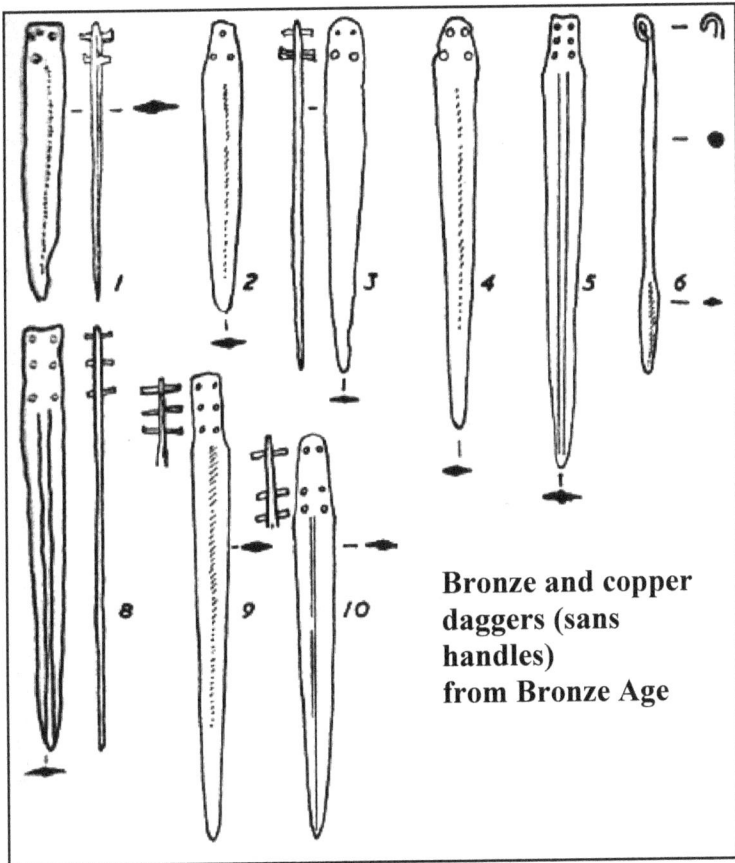

**Bronze and copper daggers (sans handles) from Bronze Age**

Since Jericho and Sodom were less than ten miles distant from one another, Sodom weapons were likely quite similar in design and materials.

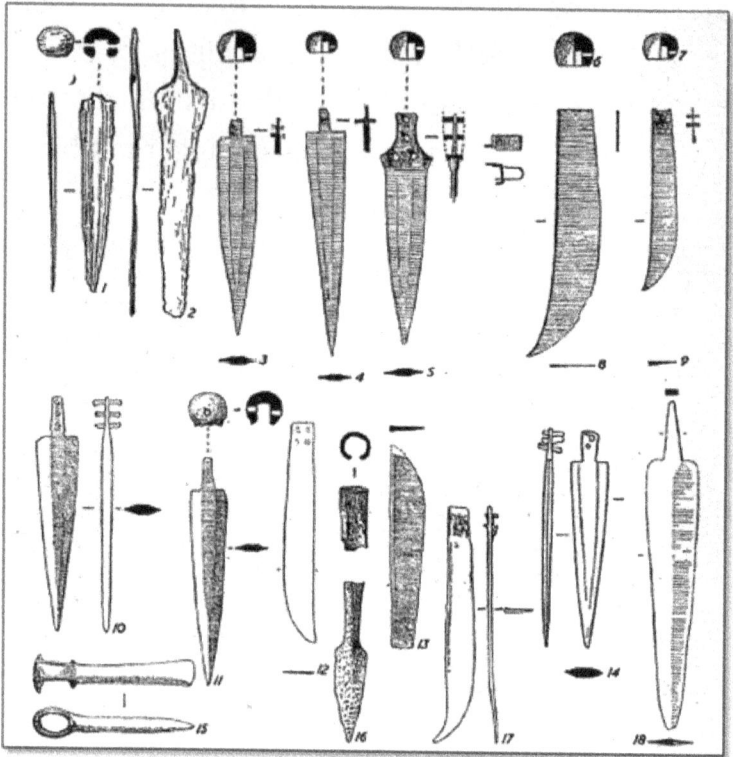

Jericho bronze and copper blade daggers from the Middle Bronze Age 2 time frame (1800 – 1540 BCE). The sophistication from the earlier types is quite striking.

The esteemed archeologist Kathleen Kenyon wrote:

*For long it was said that there was no occupation in Transjordan* [east of the Jordan River] *in the Middle Bronze Age. This would be a most unlikely state of affairs...Otherwise there must have been a prolonged intermission in occupation at a time when CisJordan* [west of the Jordon River] *was thriving and heavily populated. Such an absence of occupation is conceivable in the Negeb* [Negev], *but highly improbable in the fertile plains of western Transjordan.*[5]

Kenyon was right, although she had no knowledge that Sodom and the other Cities of the Plain had ever existed east of the Jordan River and north of the Dead Sea. But I won't rule out that she suspected as much.

Great shifts of population occurred during the Middle Bronze Age 2 time period most markedly during the period from 1800 to 1700 BCE. This was the time the Hurrian and Habiru peoples spread into the Land of Canaan. Of particular interest to us are the Habiru, for these people are thought to have included the Hebrews. These were stateless nomads, warrior bands who sometimes hired out as mercenaries. Evidence suggests they

---

[5] *Amorites and Canaanites* 1966, Oxford University Press p. 64.

were a part of, or allied to, the people the Egyptians called the Apiru. These people in fact comprised the Hyksos that conquered Lower Egypt and installed their own pharaohs for two and a half centuries.

For the most part the Hyksos passed through the Levant and into Egypt with little apparent impact on the existing culture. But the passing of these armed groups influenced a change in fortification, that of an earthen bank surmounted by a wall as opposed to a ground level free standing wall.

The excavations at Jericho demonstrate this evolution of defenses quite clearly. Walls found at lower levels of the excavation are of the free standing type, while later on the walls are built upon a sloping and plastered earth rampart rising from a stone revetment. Excavations at Sodom have revealed this same form of defensive wall combined with a plastered rampart. Sometimes a deep ditch was dug around the revetment but Kenyon says there was no ditch at Jericho. The advent of this type of elaborate defense suggests a significant increase in the town's population. Jericho, like Sodom, was now a walled city-state of the Kikkar Plain. Regarding these defenses Kenyon, in her book *Digging Up Jericho* wrote:

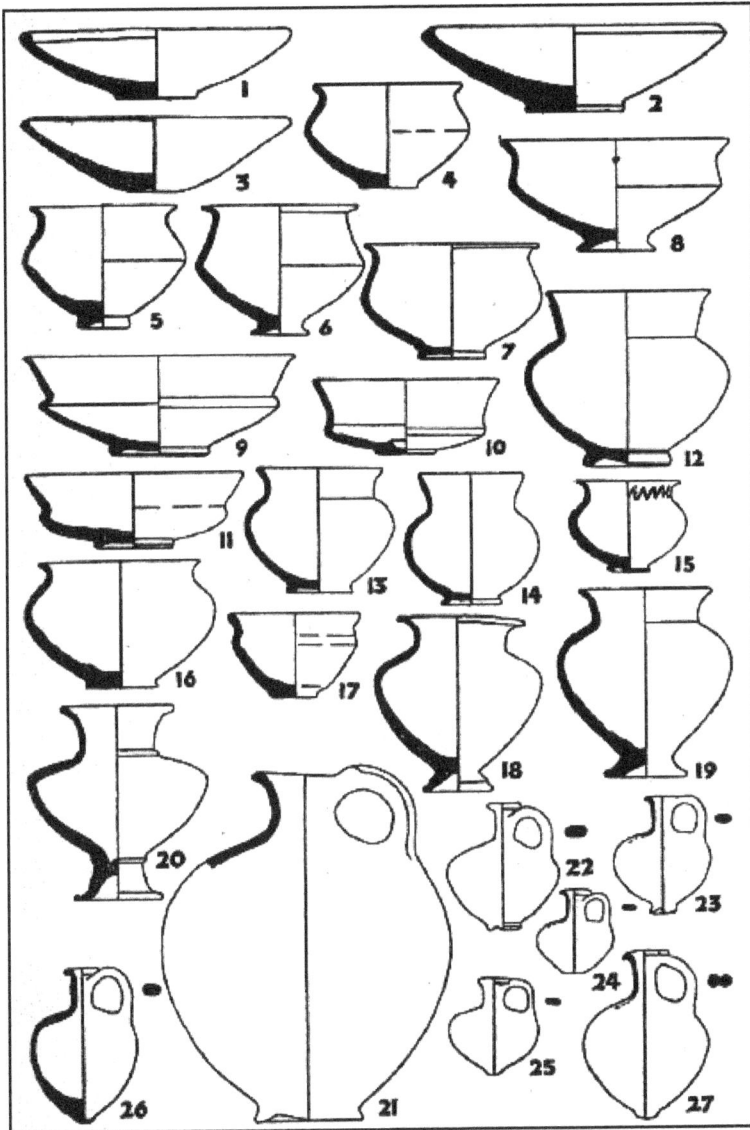

A good selection of pottery styles from Bronze Age Jericho. I fully expect that similar styles will be unearthed in future excavations at Tall el-Hammam.    Source: *Amorites and Canaanites.*

*On the north, west, and south sides...the defenses of the MBA* [middle bronze age] *may best be visualized as a wall crowning a bank with a steep plaster faced slope, with a base revetment on the outside, and a shorter slope on the inside.*

The Middle Bronze Age houses on the east side of Jericho swing out in a semi-circle to include the spring, putting a water source inside the city for the first time.

In Dame Kenyon's opinion this change was so important that she explains it even further.

*The very great interest* [in] *these defences* [sic] *is that they are so different in conception from those of the preceding period. Instead of a wall crowning an existing slope, at Jericho a slope formed by the debris of earlier occupation...they consist of a wall crowning a great artificial bank, retained at a slope considerably steeper than the natural angle of rest of the soil by a plastered surface and a massive revetment.*

A few paragraphs later she states that a similar structure is located in the Nile Delta. It would be interesting to know which of the sites was older and thus perhaps pinpointing who developed this more robust defensive wall system. Centuries later Dr. Collins found the same type of defensive wall system at Tall el-Hammam. This is another indication that Sodom and Jericho shared much in the way of culture and technology.

Cuneiform tablets, although few in number, have been found in Bronze Age excavations in Canaan including among the ruins of Jericho. Sodom was much larger in size than Jericho, and with only a small Iron Age population after its destruction, if Sodom had cuneiform tablets, they should be mostly undisturbed. If a royal or city archive of these precious written records were found in Tell al-Hammam they would greatly add to our knowledge of Sodom. Perhaps among the records could be found even the actual Canaanite name for the city the Hebrews called Sodom.

A 1933 article describes the cuneiform tablet found in Jericho.

*The tablet discovered is a small one, badly scorched, and still coated with the dust of more than thirty centuries. But even in its present condition several cuneiform signs can be distinguished on its surface.*

## The Kikkar Plain

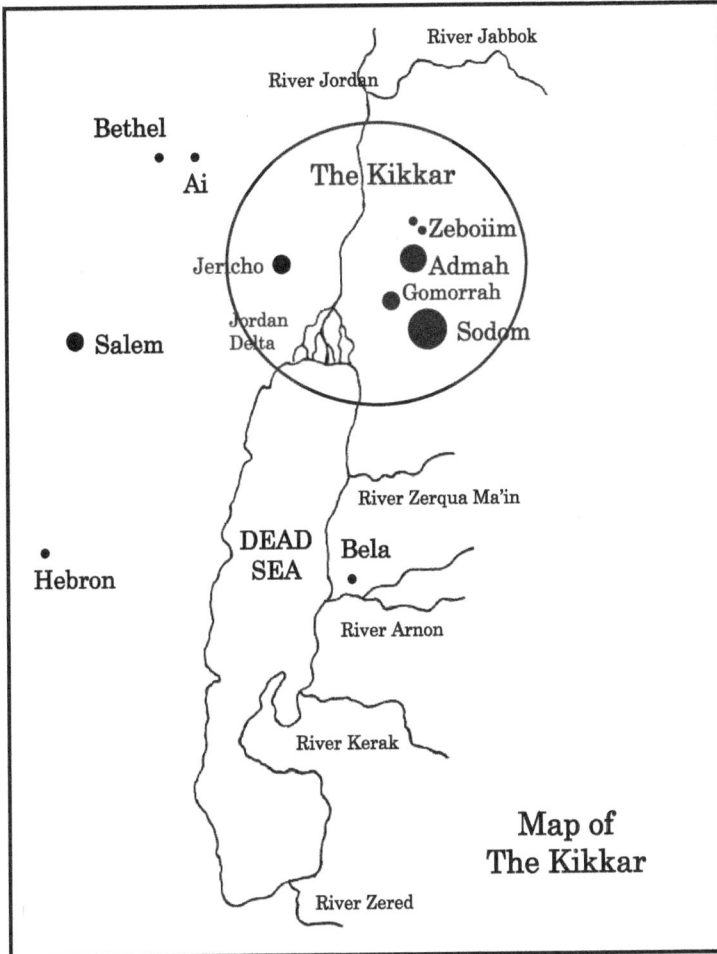

Above is a map of the Kikkar Plain (circled area) showing the Cities of the Plain and their geographic relationship to one another. Sodom, Gomorrah, Admah the twin towns of

Zeboiim and Jericho comprise the five Cities of the Plain. Adapted from *Discovering the City of Sodom.*

About 10,000 years ago Neolithic (Stone Age) farmers were living in the Jordan Rift Valley to the west of the nearby river. Their crudely constructed mud brick houses were clustered around a freshwater spring. This reliable water source encouraged their ancestors to forego their nomadic wanderings and to settle down. The valley was lush; game was plentiful, and combined with their newly adopted practice of planting cereal grains, promised a relatively secure life.

Village hunters found gazelle and wild cattle grazing the plain as well as water fowl and their eggs on the banks of the river. Fishermen armed with spears and nets also brought back fresh fish from the life giving river. The lives of the people were indeed much better, but this small village, destined to become the city of Jericho, was not the only settlement in the river valley. Others were growing on the east side of the river, and the competition for resources among these groups was heating up.

Pressure continued to rise and led to raiding parties that disputed hunting grounds and planting fields. A kind of hit and run warfare became the common practice and small skirmishes took on wider proportions with larger and larger raiding parties. This situation, coupled with nomadic peoples also raiding as they passed led the people of Jericho to take an unprecedented action. They raised a wall around their town. This was no mud brick wall, but was built of stones carried from the banks of the river a mile away. When completed, the wall stood 15 to 16 feet tall with a ditch dug around its outside perimeter.

This was something new and it even had a tower that rose about twice as high as the wall with an internal staircase of 22 steps. Here, guards had a wide view of the river valley as well as the travel routes and approaches to the town. It was an astounding sight to the other peoples of the plain, who were not far behind in copying Jericho's lead.

Granaries were constructed inside the walls and food stored in them. The town now had a protected food and water source. They ground the grain into flour for bread and fermented it to make beer. The wall gave them firm control over the west bank of the river. If the origins of our modern world can be traced back to one particular event and location, the Jordan River and the walled town of Jericho is a good candidate.

Eventually all the Cities of the Plain except for one, was situated on the east bank of the Jordan River and therefore *not* in what came to be called the Land of Canaan. This region and these cities comprised the Kikkar Valley, a circular shaped plain that encompasses that portion of the Jordon River just north of the Dead Sea. It was and still is a lush and verdant region politically divided by the Hashemite Kingdom of Jordan and the Nation of Israel.

To help in understanding the timeframe of this book and the events it describes, the following dating system, developed by Dr. Steven Collins[6], is offered for the reader's convenience and use.

| | |
|---|---|
| Chalcolithic Period (CP)[7] | 4500 to 3500 BCE |
| Early Bronze 1 (EB1) | 3500 to 3000 BCE |
| Early Bronze 2 (EB2) | 3000 to 2700 BCE |
| Early Bronze 3 (EB3) | 2700 to 2350 BCE |
| Intermediate Bronze 1 (IB1) | 2350 to 2200 BCE |
| Intermediate Bronze 2 (IB2) | 2200 to 2000 BCE |
| Middle Bronze Age 1 (MB1) | 2000 to 1800 BCE |
| Middle Bronze Age 2 (MB2) | 1800 to 1540 BCE |
| Late Bronze Age 1 (LB1) | 1540 to 1400 BCE |

---

[6] Dr. Collins is the director of the excavation project of Tall el-Hammam, the modern name of the ruin site of Bronze Age Sodom.
[7] Sometimes referred to as the Copper Age.

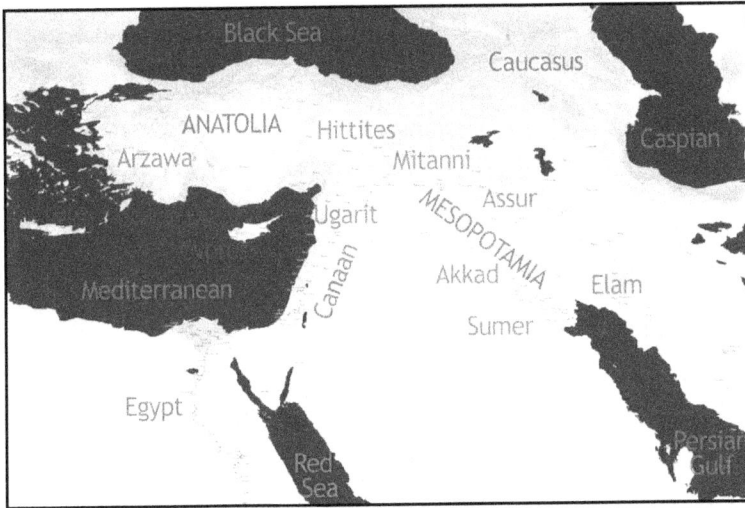

Map of the Ancient Near East. The Kikkar valley is located near the letter "C" in the word Canaan.

| | |
|---|---|
| Late Bronze Age 2 (LB2) | 1400 to 1200 BCE |
| Iron Age 1 (IA1) | 1200 to 1000 BCE |
| Iron Age 2 (IA2) | 1000 to 586 BCE |
| Iron Age 3 (IA3) | 586 to 332 BCE |
| Hellenistic Period (HP) | 332 to 63 BCE |
| Early Roman Period (ERP) | 63 BCE to 135 CE |

Painting of a Chalcolithic or Copper Age city. Courtesy Jose M. Yuste.

What's offered in the next chapter is a description of the Cities of the Plain, painting as complete a picture of each one as is possible at this time. Sodom is the only city among these that has had any advanced archeological investigation to this date except for Jericho. Even so, the surface of what lies below Tall el-Hammam[8] has only been scratched. We begin by taking a close look at the

---

[8] The modern name of the ruin of Middle Bronze Age Sodom. Tall is an Arabic word that refers to a mound built up by successive human settlements.

most famous city of the eastern Kikkar plain, the city of Sodom.

# The Cities of the Plain

## Sodom

*It was the best of times, it was the worst of times, it was the age of wisdom, it was the age of foolishness, it was the epoch of belief...it was the season of Light, it was the season of Darkness...*

To this well known opening paragraph of Charles Dickens' book *A Tale of Two Cities* I would add; *...it was the season of Life, it was the season of Death.* This is an apt description of the Middle Bronze Age for Sodom in the ancient Land of Canaan.

Most people have heard of Sodom or even more commonly the phrase Sodom and Gomorrah, for these two cities are almost always spoken of together. But there were others that archeologist Dr. Steven Collins has included by name in his book *Discovering the City of Sodom* as does of course, the book of *Genesis*.

Sodom is, without a doubt, the most famous of the Cities of the Plain, as well as the geographical and socio-political-economic center of the entire Kikkar plain. Dr. Steven Collins, director of the

Tall el-Hammam excavation project, believes that Sodom may have been founded when the people of the first large city in the Kikkar, Tulayat al-Ghassul (sometimes written Teleilat Ghassul), southwest of Sodom, lost its local water supply toward the end of the Chalcolithic (Copper Age) Period. This forced the likely large population to move to another well watered area. There, they constructed the city of Sodom.

The meaning of the name *Sodom* is unknown, but it is a Hebrew word and unlikely to be that city-state's actual name as rendered in the language of its citizens.[9] I prefer the name I derived from the Akkadian language of Uru-Dural meaning "city of the river" or perhaps just "river city". The root word Uru means city and when applied to the then city of Salem becomes Urusalem (city of peace) and eventually, Jerusalem. In like manner Urudural would eventually become Jerudural, a much better name for this ancient city than is Sodom. But alas, at least for now, that is the name the world knows this city by, so Jerudural will have to remain Sodom.

---

[9] Likely either Canaanite or Akkadian.

By far Sodom was the biggest city located upon the Jordan River and had been so for many centuries and plausibly for more than a millennium.

A Middle Bronze Age city, its history dates back quite likely into the 4th *millennium* BCE. To put it another way it was first established almost six thousand years ago. Until its destruction about the year 1700 BCE, it existed in one form or another for at least one thousand five hundred years and must be ranked among the oldest cities of humankind. To put Sodom in some historical perspective, the famous city of Troy was destroyed during the Trojan War about the year 1180 BCE, almost six hundred years after Sodom ceased to exist. Rome, that most famous of cities, was not founded until 753 BCE, a thousand years after the destruction of Sodom!

The first fortification of Sodom was a mud brick wall around the lower city that was more than 5 meters (approximately 16 feet) thick. This early wall contained gates, guard towers, an interior roadway and market plazas. Pottery sherds are available in vast quantity and allow an accurate date of its construction. This wall was constructed about 3000 BCE near the beginning of the Early

Bronze Age time frame and suggests a rise in violence, raiding, and war in the region.

The Sodom of the Middle Bronze Age was a walled and stoutly built city-state by the time of the two Hebrews known as Abram and Lot in the *Old Testament* arrived.

Sodom's king was the undisputed leader and lord of the Kikkar plain. The city's soldiers provided protection for the city that extended to its nearby daughter-city of Gomorrah. At the end of its life, Sodom had grown large and powerful, in both size and population, and was ten times larger than nearby Salem (later Jerusalem) located a short sixty-five kilometers away, with which Sodom had a political relationship. It is even possible that *Yahweh* was the accepted god of Sodom. Certainly, the goddess Asherah was worshiped and among many of the Hebrews she was considered the consort of *Yahweh.*

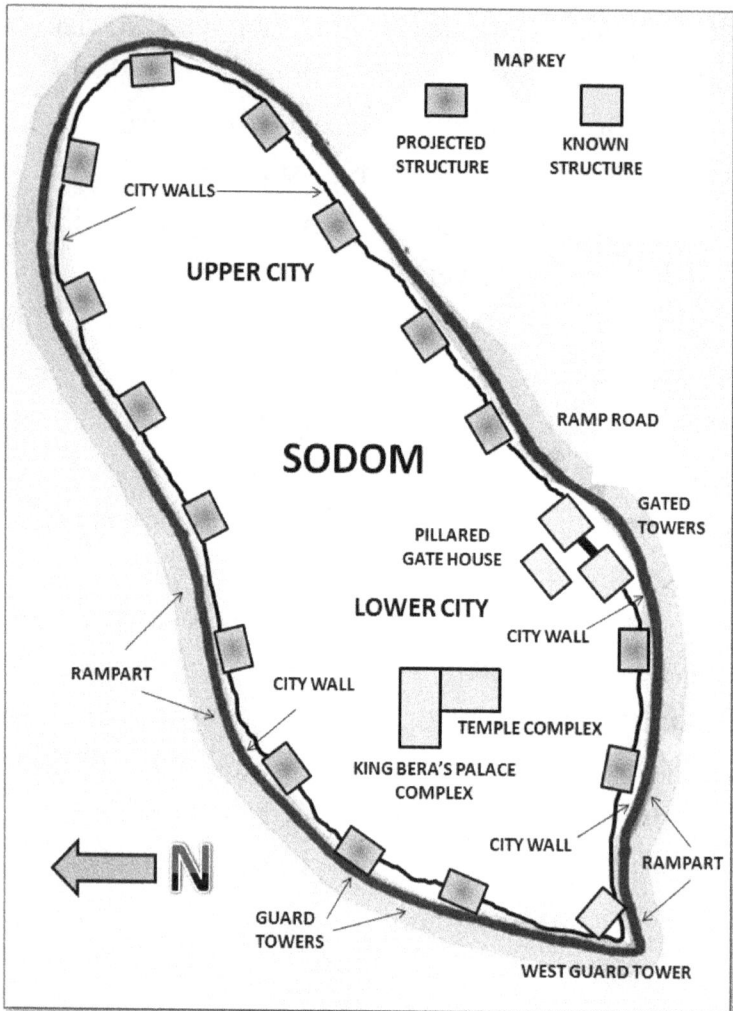

Map of the City of Sodom based on excavation drawings and notes from the Tall el-Hammam Excavation Project. A well fortified city, no archeological evidence has been found that the city was ever forcibly breached by invaders. This is an ongoing project and I expect to have to update this on-demand book as new facts are discovered by Dr. Steven Collins and his associates.

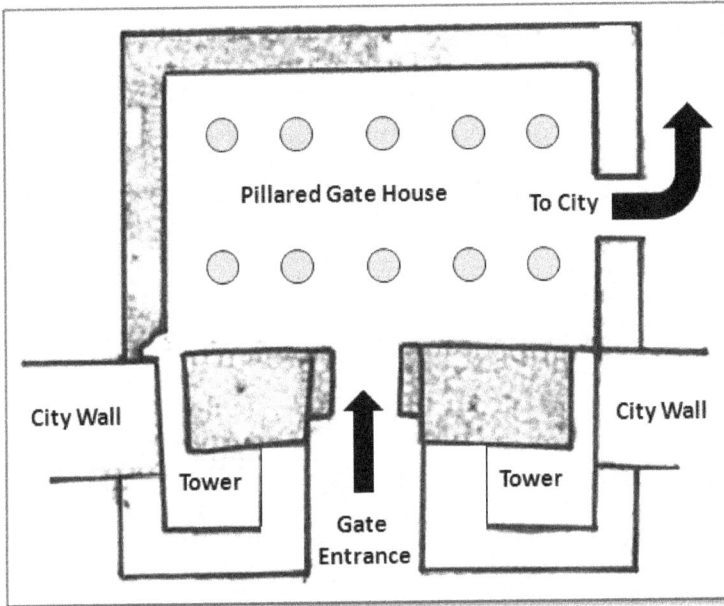

Above is a detail of Sodom's towered gate as it existed from the Middle Bronze Age to the destruction of the city (2000 to 1700 BCE). The towers contained soldiers for defense of the gate. The pillared gate house acted like a clearing station for arriving groups and individuals alike. It may have been here that Lot, an elder of the city or "judge", performed his duties as liaison between the Habiru and the citizens and administrators of the city to include King Bera.[10] This illustration is extrapolated from drawings and notes of the Tall el-Hammam Excavation Project. This is the gate and gatehouse through which the angels sent by *Yahweh* would have entered Sodom. If the angels were as striking in appearance as the *Bible* relates they would not have been able to enter Sodom without challenge. In addition to the guards, officials of the city besides Lot were no doubt on hand to observe those allowed access. These facts present a reason to believe the story of angels visiting Lot is apocryphal in nature.

Since Lot was tolerated and even became an important man in Sodom, his belief in *Yahweh* was evidently not objectionable to the citizens or to King Bera of Sodom. If Lot remained in Sodom and was there during its destruction, his remains may lie inside the gatehouse. More likely they were completely destroyed along with the city.

The walls protecting Sodom were massive, and at least fifteen feet thick. The wall itself was protected by a steep sloping glacis that provided invaders only a treacherous footing on its slick surface. The city gates, recently reduced in number to restrict access even further,[11] were

---

[10] It is quite likely that other "judges" with other duties were present as well.

[11] Some gates deemed unnecessary were filled in with dirt and debris including pottery sherds for increased security.

defended by guard towers filled with soldiers ready to respond to any threat. Archeologists have not found any indication that the city was ever penetrated forcibly by invaders. For its time it was virtually impregnable. A packed earth road circled the inside and followed along the perimeter of the walls, allowing the easy movement of people and goods.

Within these formidable walls rose the residences, palace, and temple buildings as well as at least one city plaza likely situated just inside the main city gate. At least fifty percent of all buildings in Sodom were administrative in nature. This seems to indicate it was indeed more of a capitol for all the Cities of the Plain. The plaza provided an excellent venue for an agriculture market and was most likely surrounded by merchant shops offering pottery, jewelry, precious metals, leather, and other trade goods. Two fresh water springs, located inside the city and therefore defended, provided life giving water for the citizenry and their individual gardens and trees.

The survival of Sodom during the drought preceding the arrival of Abram confirms the flow of the Jordan River and other available water sources such as springs and creeks, sustained Sodom and her close-by daughter cities.

About eighty acres in size, Sodom was located eight kilometers north of the Dead Sea's shoreline as it existed then, and twelve kilometers east of the Jordan River.

Two nearby fords likely serviced by pull-rope ferries led to roads that traveled directly west to the walled city of Jericho. The fact that there were two fords and two roads indicating heavy traffic, underlines the importance of Jericho in the trade economy of the Cities of the Plain. These were busy routes of travel, well used by caravans transporting goods to the western cities of Jericho, Bethel, Salem, Hebron and beyond. Caravans of camels[12], donkeys and horses passed through the Jordan and Siddim valleys and past the various cities to even more exotic destinations to the north along the sea and south to Egypt. This constant movement of trade brought increased commerce and riches to all the cities of the Kikkar with Sodom chief among them. This was indeed a thriving, bustling civilization that without a doubt was comprised of a hundred thousand people and possibly many more.

Scholars are unsure whether the Jordan River was navigable or not. I understand that, the river

---

[12] There is some debate among researchers as to whether camels were domesticated sometime after this time.

today is narrow and shallow and it's hard to visualize boats of any real size upon it. However, things could have been much different three thousand eight hundred years ago. There is little doubt however, that the river was fished to provide another source of protein for the people. These fish, taken fresh from the river supplemented the salted fish brought to the city from the Sea of Galilee well to the north.

The King of Sodom is called Bera in the *Old Testament*. This was no doubt not his name, or was at least a very unfortunate rendering of it in the language of those who wrote the *Old Testament*. For in Hebrew Bera translates as *Son of Evil*. At the time Sodom existed many of the languages of Canaan were similar, and with but one additional letter, the letter *h* added to the end of the king's name, becomes Berah and translates as the *King of Blessing* which sounds much more plausible to me. If we accept this assumption as correct, the writers of the *Bible* many years later wanted to underscore the evilness they ascribed to Sodom and eliminated this one letter putting the desired spin on the king's name.

The archeology suggests that the king was an able administrator and for the most part led a prosperous and populous city and region of ancient times. In addition to being a crowned king, Bera was the recognized lord of the Kikkar Plain.

The idea of Sodom's evilness may come from the god and goddess that Sodom and the other cities of the plain likely worshipped (as well as many other peoples in Canaan). Quite possibly in the city's temple stood large stone figures of Shamash, god of the Sun; and the goddess of the moon with the unfortunate name of, Sin.[13] The people likely had miniatures of these figures in their homes and places of business and without doubt prayed to them as believers.

In a similar fashion as Melchizedek of Salem, who was the high priest of *Yahweh* as well as king of the city, King Bera may have had some priestly functions to perform in the worship of his city's deities.

Today, Sodom is a ruin, many stories high and known by the modern name of Tall el-Hammam. It is under excavation by Dr. Steven Collins who deserves much credit for discovering this important city's location. The future work of Dr. Collins and other archeologists will provide new facts about this fascinating place and its people in the future. It will also be important in another way. As the stones, bricks and debris are shifted

---

[13] This is but one of the names used for the goddess of the moon.

under the archeologist's careful spade, artifacts will come forward to highlight not only the history of Sodom, but just as and possibly more important, the secrets of its destruction. For in fact what Dr. Collins, his colleagues, and their many volunteers are researching is not only the past, but quite possibly the very near future.

A close-up of the 6th century Madaba mosaic map. The two unnamed cities are in the position where the ruins of Sodom and Gomorrah are today.

## Gomorrah

Like many of the other cities, the meaning of this city-state's name is lost in the mists of time. However there may be a reference to the spring

47

inundations of the Jordan River within the name. This may be because of the location of Gomorrah, for of all the cities of the time, it is closest to the Jordan River. In the *Bible* Gomorrah is referred to as the daughter-city of Sodom. The term daughter-city implies that Gomorrah was a direct offshoot of Sodom, settled by people who came from Sodom and augmented over time by residents from other places, but remaining closely tied to Sodom throughout its existence. It is quite plausible that the king of Gomorrah, known as Birsha (in *Genesis*) or Barsus (according to the historian Flavius Josephus[14]) was a close relative of King Bera of Sodom, perhaps even his son.

Among the Cities of the Plain, Gomorrah seems to be the third largest city behind Sodom and nearby Admah and perhaps on a par or slightly bigger than Jericho. Gomorrah too was a walled city with in-depth defenses, but likely relied on Sodom and its larger garrison of soldiers for its protection during particularly dangerous times. A short four kilometers west and north of Sodom, it was a bit closer to the Jordan and surrounded by many hamlets, small settlements, holdings, and

---

[14] Titus Flavius Josephus (37 – c. 100) was a first-century Romano-Jewish historian who was born in Jerusalem—then part of Roman Judea—to a father of priestly descent and a mother who claimed royal Jewish ancestry.

farms. It did a considerable business in lucrative trade. Caravans going to Sodom no doubt paused at Gomorrah to conduct bartering, and to seek shelter and refreshment from the public houses (inns) of the city.

Without a doubt there was also a rich and thriving trade in the oldest profession among the women of Gomorrah. This was a common practice and by no means unique to Gomorrah, Sodom, or any of the other cities. The Akkadian word for such women is harimtu and calls to mind the word harem that has a somewhat different modern meaning. The sacred prostitutes of the temples of Canaan were called qadishtu and were highly regarded socially. The use of prostitutes in support of religious sites was not an unusual practice in the Ancient Near East and can be found among the Greeks and other ancient societies as well.

As a walled city with its own king, Gomorrah had a contingent of soldiers who likely were as loyal to Sodom as they were their own city-state. In fact the two cities were so closely united they were thought of as almost one entity, and so the commonly used phrase with the two almost always connected is *Sodom and Gomorrah.*

The ruin of Gomorrah is known, one of the many smaller ruins scattered around the modern ruin called Tall el-Hammam. Future excavations will better tell us which ruin is in fact Gomorrah, and fill in its history and importance to the alliance of the Cities of the Plain.

The *Bible* lists Gomorrah along with Sodom as a city raided during the invasion of the Eastern Kings under King Chedorlaomer of Elam.

## Admah

This city-state's name is a reference to red earth, or red clay and may have been a center for the production of pottery. Future investigation of its ruin may give clues as to why it bore this designation. Similar in size and somewhat larger than Gomorrah, Admah likely had the most sway in the councils of the Cities of the Plain besides Sodom. The king of Admah was Shinab which seems to have the meaning of *One who Draws Money*. This likely refers to his duties as a tax collector, as indeed all kings were, it could also mean that he was in charge of the coalition of the Cities of the Plains' treasury or at least its accounting. However this last possibility presupposes a close confederation of the cities.

Perhaps excavations of the ruin of Admah may clear up this question.

An alternative name for this king or perhaps a reference to a succession of Admah kings is Sumobor, and possibly others. An alternative meaning of Shinab is *Splendor of the Father* which seems a more traditional meaning for a man's name.

Also like Gomorrah, Admah was walled and had its own city-state army garrison. As with the other Cities of the Plain, Admah was committed to its alliance with the leading city of Sodom. Admah too flourished and grew rich amid the fertile fields of the Kikkar plain. It is not listed in the *Bible* as one of the cities raided during King Chedorlaomer's invasion of the Kikkar Valley.

## Zeboiim

The king of Zeboiim is listed as Shemeber in *Genesis* which translates into *The Illustrious*, or the alternative meaning of *One who Soars High*.

The name Zeboiim is a reference to gazelles, and is assigned to twin towns, both about the same size and population to one another, likely smaller in size than Gomorrah, but together large enough to warrant their own king. The double *i* in the name is an indication of plurality. They could have been known for example, and only an example, as east Zeboiim and west Zeboiim and may have stood on either side of the trade route coming into the Kikkar from the north. It is known they stood on either side of a small stream.

In effect, the position of the Zeboiim towns suggests they were guardian settlements and may in fact have been constructed as army garrisons with the mission of defending the northern approaches to the Kikkar. Admittedly speculative at this time, this arrangement makes sense, as the most likely invaders would come from the north where empires were under construction and looking to add to their territories and tributary populations. And so the name may mean, besides the reference to gazelles, a reference to speed, as in the soldiers of Zeboiim were in modern terms, a ready reaction force. The unit itself may have had gazelle in its name or insignia. This would be similar to the King of Elam's elite force, his palace guard, known as "Bears (or bearers) of the Silver Spears".

A sizable military contingent stationed here could cause an invader to seek another, less well guarded approach and as we will see later, that is exactly what happened. If they were army towns, then the soldiers would have supplemented the rations provided to them from the Kikkar confederation by working at their own gardens along with their wives, servants and other dependents.

Some may think that this timeframe could be too early for the concept of standing armies. However, half a century before Sodom's fall the Hyksos, thought to be a people from Canaan, invaded and conquered the northern portion of Egypt and successfully ruled it for more than two centuries. If you wanted your city-state or empire to survive, you needed a ready, and admittedly expensive, standing army.

Soldiers gear at the time was kept simple and likely consisted of sturdy foot wear, a basic garment somewhat like a kilt with a form of tunic optional in the cool season and a light blanket that also doubled as a pack to hold rations. His arms included a bronze tipped spear, large bronze knife, or possibly an ax, a lightweight shield and perhaps a stone throwing sling.[15] Heavier shields and even

---

[15] Researchers have shown that a sling in the hands of a well practiced soldier could deliver a stone or lead pellet with the equivalent impact of a modern .45 caliber pistol round. The sling was indeed a quite potent weapon. Some slingers wore the sling around their head while marching.

armor may have been available should a campaign require them.

Special formations of slingers and/or archers might also have existed. Their training, when they trained, was in hand to hand combat and developing efficiency with the knife, spear, and sling or bow. Armies were made up of mostly infantry, with some of the richer nations or empires able to field a number of chariot mounted warriors or some cavalry, but these troops were extremely expensive and the exception rather than the rule. The coalition may have fielded chariots but if so, they were few in number and likely used to move the kings and their generals around the battlefield.

If the soldiers wore any armor it was probably made of leather or layered garments. The sword would later develop out of the long dagger-like bronze knife or perhaps from an agricultural implement.

The army of the coalition was made up by this time of actual soldiers, who marched together, trained together and fought as a unit with their own command structure. This arrangement was superior to the warriors of the nomadic bands who fought as individuals for the most part. But as will be seen, when the army did fail the cities, it was warriors under the Patriarch Abram, with their unconventional style that would save the day.

## Bela

This town is also known as Zoar and was likely located on the southern edge of the plain and hard upon the river Arnon east of the Dead Sea. Zoar means *small* and subsequently Bela was the smallest of the cities, possibly near the size of one of the towns of Zeboiim. It may have even been a tent city of Bedouin-like nomads nominally loyal to Sodom.

Bela too seems to be a guardian town, as it sat astride the trade route leading from the south into the Kikkar. Although a functioning community in its own right, Bela was likely more of a forward observation post for Sodom and its coalition rather than an actual impediment to an invading army. The king of Bela is not named in Genesis or by Josephus and appears only once in the chronicles of Theophilus[16] who lists his name as Balach. This name is possibly apocryphal (as indeed are all the king's names) and used only for reference in

---

[16] *Theophilus* is the name or honorary title of the person to whom the Gospel of Luke and the Acts of the Apostles are addressed.

Theophilus' writings. Other indications are that the name means boy, or youthful man.

Bela provided ready access to the pits of tar near the Valley of Siddim, dug and used in commerce and in some kinds of construction. Beyond the Jordan delta and on a drier southern edge of the plain, Bela nestled in the palms of an oasis watered by sweet rivulets and the river Arnon flowing from the Mountains of Moab to its east. The dates grown among its palms were the sweetest to be found, they were quite desirable, and the town did a brisk trade in them.

A number of researchers, including many archeologists, think that Bela is the fifth city of the plain, but in fact it is located outside the Kikkar Plain. There is only one city that can be the fifth city of the plain and that is the city of Jericho, equally as famous as Sodom and Gomorrah.

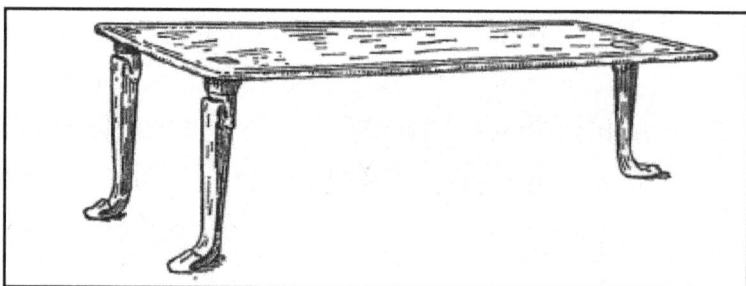

Middle Bronze Age three legged table from Jericho. An almost universal item of furniture, the table was long and narrow and because Jericho was built on a slope, it was designed to sit on the uneven floor.

## Jericho

I include in this account one more city of the plain not normally considered among those destroyed when *Yahweh* wreaked his destruction, the walled and well fortified city of Jericho.

Because ancient Jericho has been excavated to a large degree, there is more known about it and its inhabitants. In addition, being located very near Sodom, its cultural styles such as pottery and building construction may give us glimpses of those of Sodom. I will address this city more fully than the others due to the rich information written about it, especially by archeologist Kathleen Kenyon, a recognized expert on Jericho's history.

This ancient town is unique among the Cities of the Plain because Jericho is quite likely the oldest continuously[17] occupied city on Earth.

Archeologists have found evidence of settlement in the area of Jericho beginning 10,000 years

---

[17] This designation is accepted even though Jericho was abandoned for a time during the Middle Bronze Age. The reason for the abandonment is addressed later in this narrative.

BCE that is, 12,000 years ago. In the latter half of the Middle Bronze Age (MB2) Jericho, along with the rest of the Jordanic Civilization entered a period of prosperity. During this phase of its existence its citizens widened and strengthened its walls and fortifications.

Most researchers do not include Jericho in the destruction of the Cities of the Plain and there is no mention of its destruction by *Yahweh*'s hand in the *Bible*. But I include it for the following reasons.

None of the Hebrews under the leadership of Abram and including Abram ever set eyes on Jericho after the destruction of Sodom. All indications are that Abram and his people fled precipitously far to the south to the city-state called Gerar and ruled by King Abimelech. Afterward, there seems to be no further communication between Abram and King Melchizedek of Salem, the high priest of *Yahweh*. Its clear Abram had no knowledge of the destruction of Jericho, so its demise was not verbally handed down and therefore did not make its way into the *Bible*. However, destroyed it was.

Geographically Jericho is located on the Kikkar Plain; it was one of the Cities of the Plain. Jericho is twice as far from Salem as from Sodom, 15 miles as opposed to 7.5 miles. Sodom was undoubtedly the dominate city of the region, including what is known as the Land of Canaan, for it was ten times larger than Salem and Jericho. There can be little doubt but that Jericho looked to the fertile lands of the Kikkar and its cities, in particular Sodom; politically, militarily, culturally, and economically. There was without much doubt blood relations among both populations, perhaps even between the two rulers.

A sizable marching army could move from the east side of the Kikkar, cross the Jordan River, and arrive at Jericho in less than half a day's march using the existing fords and roads. Like the other Cities of the Plain, Jericho was walled and had its own military garrison. It prospered along with the other members of the coalition from the trade that passed by its gate via a major east-west road on the way to and from Sodom and the rest of the Kikkar, thus making it a strategic crossroads.

The large fertile plain of the western Kikkar watered by a nearby gushing spring, earned Jericho the nickname *The City of Palms.* Jericho sits 670 feet below sea level and about 3,000 feet in elevation below Salem.

Jericho in the Canaanite language means *The Moon* and it was indeed the seat of moon worship. The name of the king of Jericho at the time of its destruction is not mentioned in *Genesis*, but the name of its legendary, and perhaps its first king is King Keret. There is archeological evidence that Jericho is a very ancient city, likely even older than Sodom.

The archeology of Jericho is almost as muddied as the myriad investigations of the assassination of President John F. Kennedy. Personally, I think the most accurate of those that have investigated Jericho is archeologist Dame Kathleen Kenyon. She had no personal agenda in the excavation unlike some who later came seeking to vindicate their own views or that of others funding their research.

From 1952 to 1958 the ruins of Jericho were excavated and interpreted by Dame Kathleen Kenyon, the recognized authority on the ruin of Jericho. Kenyon dated the destruction of Jericho in the Middle Bronze Age in keeping with the same rough dating as the destruction of the other Cities of the Plain. However, her dates do place its end at about 1550 BCE. She is close to the more correct date of about 1700 BCE, and the comparison of artifacts found at Jericho with those being excavated by Dr. Steven Collins in the ruins of Sodom[18], should sharpen up these dates and I am confident they will agree with the end date of the other Cities of the Plain.

At the time of its destruction Jericho was not at ground zero like Sodom and Gomorrah and their allies, but the damage dealt the city was massive. There is little doubt that walls were thrown down and people killed in their hundreds if not thousands. This would be especially true of the eastern half of the city, where a shock wave, followed almost immediately by a thermal wave that ignited anything of wood or flesh exposed to it, did the worst damage. Dame Kenyon wrote:

*The destruction was complete. Walls and floors were blackened or reddened by fire, and every room was filled with fallen bricks, timbers, and household utensils; in most rooms the fallen debris was heavily burnt.*

Portions of the north wall were left standing, but everywhere else the wall fell from the blast or crumpled from the heat. This is in keeping with the fact that Jericho is located northwest of Sodom and ground zero. It follows that the north wall and the houses built against it would be spared the full impact of the shock wave and the heat. Kenyon

---

[18] Tall el-Hammam.

noted the walls collapsed before the city burned and that is in keeping with the effects described during *Yahweh's* crushing of Sodom.

The city did not survive the raging fires and like the other Cities of the Plain, Jericho perished. However, in stark contrast to the other cities, Jericho would rise again during the Bronze Age, to the evident chagrin of *Yahweh* who, almost four hundred years later, would send Joshua the Israelite to complete its destruction. By that time Jericho was inhabited by Amorites.

Kenyon observed:

*The final end of the Early Bronze Age civilization came with catastrophic completeness. The last of the Early Bronze Age walls of Jericho was built in a great hurry, using old and broken bricks, and was probably not completed when it was destroyed by fire. Little or none of the town inside the walls has survived subsequent denudation, but it was probably completely destroyed, for all the finds show that there was an absolute break, and that a new people took the place of the earlier inhabitants. Every town in Palestine that has so far been investigated shows the same break [my emphasis]. The newcomers were nomads, not interested in town life, and they so completely drove out or absorbed the old population, perhaps already weakened and decadent, that all traces of the Early Bronze Age civilization disappeared.*

After its destruction Jericho was abandoned for a time just like Sodom would be. But unlike Sodom, Jericho was reoccupied during the Late Bronze Age.[19]

In reference to Joshua's defeat of the City of Jericho hundreds of the year later Kenyon wrote:

*"...we have nowhere been able to prove the survival of walls of the Late Bronze Age, that is to say of [to?] the period of Joshua".*

This is likely because the people, who reoccupied the still devastated site of Jericho, built no walls. Like the rest of the Cities of the Plain, it was so destroyed it was not deemed worth the effort of rebuilding the city walls.

---

[19] *Amorites and Canaanites* p. 74.

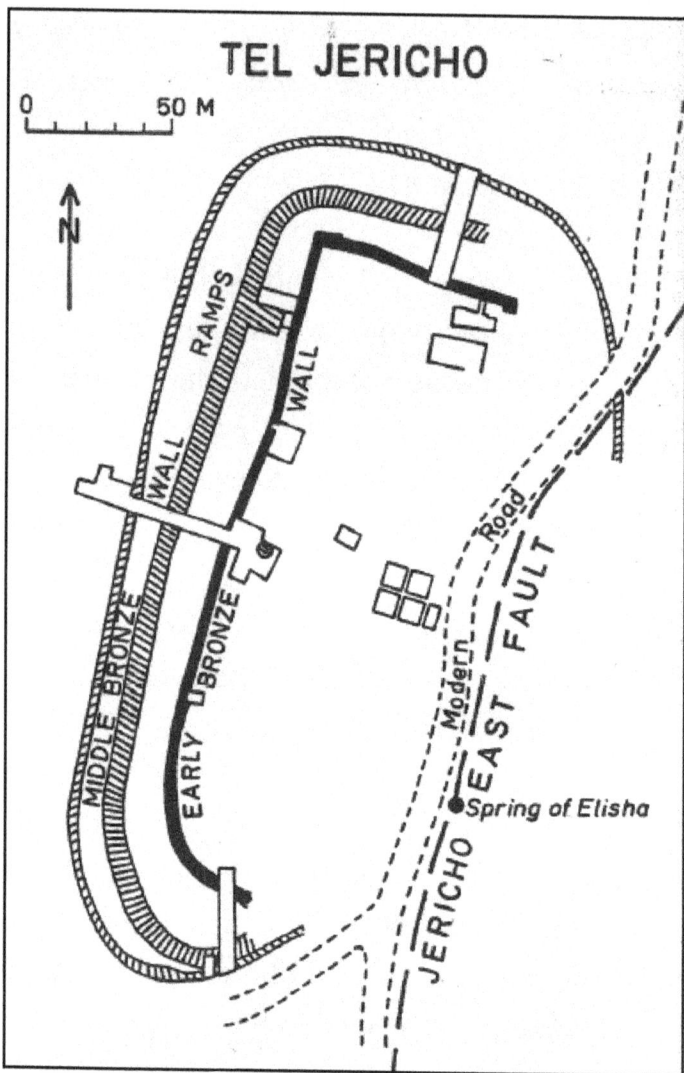

A sketch map of Tell Jericho showing the fault line to its east. Adapted from *Amorites and Canaanites.*

## Kings of the Kikkar Plain

| Source | Name | City |
| --- | --- | --- |
| *Genesis* | King Bera | Sodom |
| | King Birsha | Gomorrah |
| | King Shinab | Admah |
| | King Shemeber | Zeboiim |
| | Not listed | Bela |
| | | |
| Flavius Josephus | King Ballas | Sodom |
| | King Barsas | Gomorrah |
| | King Sumobor | Admah |
| | King Senabar | Zeboiim |
| | Not Listed | Bela |
| | | |
| Theophilus | King Ballas | Sodom |
| | King Barsas | Gomorrah |
| | King Senaor | Adamah (sic) |
| | King Synobar | Seboim (sic) |
| | King Balach | Segor (Bela) |

## The Eastern Kings

| Source | Name | City/State |
| --- | --- | --- |
| *Genesis* | King Chedorlaomer | Elam[20] |
| | King Amraphel | Shinar |
| | King Arioch | Ellasar |
| | King Tidal | Nations |
| | | |
| Flavius Josephus | King Chodolaomer | Elam |
| | King Amraphel | Shinar |
| | King Arioch | Ellasar |
| | King Tidal | Goiim |
| | | |
| Theophilus | King Chodollonagor | Assyria |

King Tidal is also known as the King of the Hittites from the west, while King Amraphel is thought to actually be the famous lawgiver, King Hammurabi of Babylon.

Today the ruins of all the Cities of the Plain except Jericho are in the Hashemite Kingdom of Jordan. All have modern names that begin with *tall* which means hill. What follows is a list of these talls and their corresponding *biblical* names. There are many more talls surrounding Sodom

---

[20] Possibly Assyria.

than can be accounted for among the known Cities of the Plain. That leaves a lot of room for expansion of the cities to future excavations.

## Talls of the Kikkar Plain

Tall el-Hammam – Sodom
Tall Kefrein – Gomorrah
Tall Nimrin – Admah
Tall Bleibel and Tall Mustah – Zeboiim.

There are other talls in the plain whose names and dates of existence are unknown.

Tall Tahuria (located north of Sodom)
Tall Iktanu (located south of Sodom)
Tall Zzeimeh (located south of Ikantu)
Tall Rama (located west of Sodom)
Tall Mwais (located south-west of Sodom)

Preliminary investigation has indicated these talls may be the remains of other Middle Bronze Age 2 cities that were destroyed at the same time as all the others. If this is proven through archeological excavation, the importance of the Jordanic Civilization may be much more than is already apparent.

If they are contemporary to Middle Bronze Age 2 Sodom, their likely inhabitants were Canaanites.

# Chapter IV
## The Eastern Kings Conquer the Transjordan

This portion of the *Genesis* account is in reference to the city states of the Transjordan area lying east, south and north of Sodom.

*Twelve years they* [the cities of both the Kikkar plain and the Transjordan] *served Chedorlaomer, and in the thirteenth year they rebelled.*

*And in the fourteenth year came Chedorlaomer, and the kings that were with him, and smote the Rephaims in Ashteroth Karnaim, and the Zuzims in Ham, and the Emims in Shaveh Kiriathaim.*

*And the Horites in their mount Seir, unto Elparan, which is by the wilderness.*

*And they returned and came to the Enmishpat, which is Kadesh, and smote all the country of the Amalekites and also the Amorites, that dwelt in Hazezontamar.*

The first people to come under attack were the Raphaims in the city-state of Ashteroth Karnaim (literally the name of the Canaanite fertility goddess Ashteroth in the Horns of the Mountains),

69

located well north and east of the Kikkar. The name of this people means "giants" and is a reference to a tribe of the Amorites, known for their very large and fearsome warriors. But despite their great ferocity, the capital city fell to the swords, chariots and the Bears of the Silver Spears of the four allied kings.

Next on the list was the tribe of people known as Zuzim meaning roamers or wanderers, a people very similar to the Habiru/Hebrew. These people lived in the Land of Ham to the east of the Jordan River. They too were brought under Chedor-laomer's rule by conquest.

The Emites, another branch of the Amorite people originally from the west of Syria, were targeted. The name has been translated variously as "the dreaded ones", or "horror" and even "terror". Regardless of their fearful name, they suffered the conqueror's heel on their necks. They too were often referred to as giants.

The next to fall in this continuing domino of nations was that of the Horites, a Canaanite people who lived in the vicinity of Mount Seir, east of the Jordan River. Their defeat by the Eastern Kings is the first reference to this people. Many years later according to *Genesis 36*, the Horites intermarried with the descendants of the Hebrew Esau, son of

Isaac and grandson of Abram. Esau's descendants became rulers in Edom and were known as dukes and later kings. It was this people that would not allow the Israelites under Moses to enter their kingdom during their exodus from Egypt to the Promised Land. Obviously the officials of Edom viewed the doubtless motley looking[21] Israelites as a threat.

Later the rabbis of Israel decreed the Edomites and Moabites were the offspring of Lot's incestuous affair with his daughters. An insult that the Israelites perpetuated by putting it in their sacred writings, including the *Bible*.

Reluctantly and flushed in anger Moses' people skirted their cousin's Kingdom of Edom and traveled to the east all the way around Edom, yet bumped into the Kingdom of Moab further north. Archeologist Nelson Glueck wrote of this confrontation that took place during the Iron Age.

*The archaeological survey of Edom soon revealed why it was that not without permission might a foreign group enter its territory. The permission refused, the applicants for entry must perforce turn aside as the Israelites were*

---

[21] They had just arrived after a long journey across the dessert.

*compelled to do. Strong fortresses barred the way on all the frontiers of Edom, and of Moab to its north.*[22]

Unable to enter Moab, the Israelites now traveled north, the only direction open to them, and crossed the Jordan River, in view of, and moving past, the ruins of Sodom and Gomorrah. Angered at the treatment by their distant relatives and spoiling for a fight, the Israelites fell upon Jericho with a vengeance.

After being conquered by King Chedorlaomer and his allied kings, the many city states of the Transjordan no doubt agreed to renew their tribute and went to work rebuilding their damaged cities and replenishing their looted treasury. It would all prove to be in vain a few bitter years later. Glueck places the end of this branch of the Jordanic Civilization at about 1750 BCE. It is not a coincidence that this ending date is so near that of Sodom, Gomorrah, and the other Cities of the Plain.

---

[22] *The Other Side of the Jordan* p. 128.

# Chapter V
# The Eastern Kings Invade the Kikkar

There were events that occurred in and around Sodom and the Cities of the Plain mentioned in *Genesis* besides their destruction. When reading the following passages, it should be remembered that long before humans developed the skills of reading and writing, history was an oral tradition. And even when writing became available, the common people continued to keep their history by storytelling.

These stories were told around the fire, likely at night as the day was spent in the pursuit of their livelihood. The stories changed, evolving with each retelling as the storyteller added or removed some detail. These stories were not just the passing on of facts about the past, but were also a form of entertainment. The teller did not want his audience to fall asleep and let some predator or enemy fall upon them in surprise so the teller made his story interesting with exaggerated claims over the literal truth, but often keeping the kernel of fact intact. In the telling of the story they remade the story.

Soldiers of King Chedorlaomer and his allies are depicted on the top line of this illustration. The soldiers of the Kikkar are shown on the bottom line. Note the eight pointed star on their shields that harkens back to the wall paintings of Tulayat al-Ghassul, Sodom's mother city. It is likely that most soldiers of the time were similarly equipped, but these depictions, especially the soldiers of Sodom, are speculative.

Only later when reading and writing became more common, did they become codified on scrolls and once written, passed down to each succeeding generation. The hero of the following passages, Abram, lived in a time, about 1760 BCE, well before the common use of the written word. Only many, many years later were these stories written down.[23] None the less, they do comprise a history.

The invasion of the Kikkar and other kingdoms of the southern Levant are addressed in *Genesis.*

---

[23] Many researchers believe these stories were first written down as long as a thousand years later.

*Genesis*, Chapter 14:

*And it came to pass in the days of Amraphel king of Shinar, Arioch king of Ellasar, Chedorlaomer king of Elam, and Tidal king of nations;*

*2 That these made war with Bera king of Sodom, and with Birsha king of Gomorrah, Shinab king of Admah, and Shemeber king of Zeboiim, and the king of Bela, which is Zoar.* [The kings and Cities of the Plain].

*3 All these were joined together* [joined in battle] *in the vale [*valley*] of Siddim, which is the Salt Sea[24].*

*4 Twelve years they* [the cities of both the Kikkar plain and Transjordan] *served Chedorlaomer, and in the thirteenth year they rebelled.*

*5 And in the fourteenth year came Chedorlaomer, and the kings that were with him, and smote the Rephaims in Ashteroth Karnaim, and the Zuzims in Ham, and the Emims in Shaveh Kiriathaim.*

---

[24] In my opinion this is a reference to a valley near the Dead Sea, not the sea itself, which geographically predates this battle by hundreds of thousands of years and perhaps as much as a million years.

*6 And the Horites in their mount Seir, unto Elparan, which is by the wilderness.*

*7 And they returned and came to the Enmishpat, which is Kadesh, and smote all the country of the Amalekites and also the Amorites, that dwelt in Hazezontamar.*

*8 And there went out the king of Sodom, and the king of Gomorrah and the king of Admah, and the king of Bela (the same is Zoar) and they joined battle with them in the vale of Siddim;*

*9 With Chedorlaomer the king of Elam and with Tidal king of nations and Amraphel king of Shinar, and Arioch king of Ellasar; four kings with* [against] *five.*

*10 And the vale of Siddim was full of slimepits; and the kings of Sodom and Gomorrah fled, and fell there and they that remained fled to the mountain.*

*11 And they took all the goods of Sodom and Gomorrah, and all their victuals, and went their way.*

*12 And they took Lot, Abram's brother's son, who dwelt in Sodom, and his goods, and departed.*

*13 And their came one that had escaped, and told Abram the Hebrew; for he dwelt in the plain of Mamre the Amorite, brother of Eshcol, and brother of Aner: and these were confederate with Abram.*

*14 And when Abram heard that his brother was taken captive, he armed his trained servants, born in his own house, three hundred and eighteen, and pursued them unto Dan* [near Damascus, Syria].

*15 And he divided himself against them,* [meaning he attacked from two directions] *he and his servants, by night, and smote them, and pursued them unto Hobah, which is on the left hand of Damascus.*

*16 And he brought back all the goods, and also brought again his brother Lot, and his goods, and the women also, and the people.*

*17 And the king of Sodom went out to meet him after his return from the slaughter of Chedorlaomer, and of the kings that were with him, at the valley of Shaveh, which is the king's dale.*

*18 And Melchizedek king of Salem brought forth bread and wine; and he was the priest of the most high God.*

*19 And he blessed him, and said, Blessed be Abram of the most high God, possessor of heaven and earth:*

*20 And blessed be the most high God, which hath delivered thine enemies into thy hand. And he* [Abram] *gave him* [the king of Salem] *tithes of all.*

*21 And the king of Sodom said unto Abram, Give me the persons, and take the goods to thyself.*

*22 And Abram said to the king of Sodom. I have lift up mine hand unto the Lord, the most high God, and the possessor of heaven and earth.*

*23 That I will not take from a thread even to a shoelatchet, and that I will not take any thing that is thine, lest thou shouldest say, I have made Abram rich:*

*24 Save only that which the young men have eaten, and the portion of the men which went with me, Aner Eshcol and Mamre; let them take their portion.*

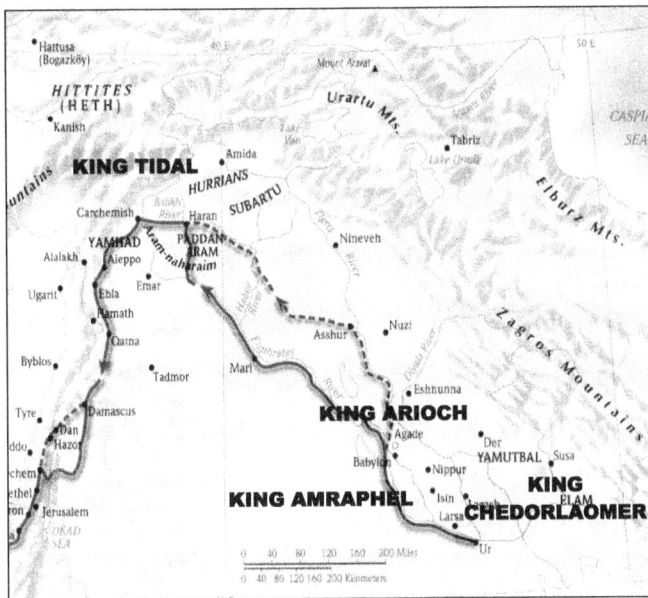

The above map shows the route Abram and his family took to reach the Land of Canaan. It also shows where the four allied kings who came to punish Sodom and the Cities of the Plain, came from. There is a big difference between these kings and the kings of the Cities of the Plain. The four allied kings were rulers of large areas or kingdoms, while the kings of the Kikkar ruled much smaller city-states. The king of Sodom, also the recognized Lord of the Kikkar, is the only one of the five kings of the Kikkar to be nearly the peer of the allied kings that came to punish him.

In his book *Discovering the City of Sodom* Dr. Collins notes that during this historical time period written agreements between geographically distant kings was known. It is evident that such an agreement was made between either the king of Sodom in his role as Lord of the Kikkar or collectively by all the kings of the Cities of the Plain with the king of Elam. It is my opinion that the former is more likely with Sodom taking the initiative in both making the agreement and thirteen years later, leading the rebellion.

For twelve years the Lord of the Kikkar sent tribute caravans loaded with goods and treasure north and east to the king of Elam. In the thirteenth year something changed. There was a wide-spread rebellion against Elam as witnessed by the fact that King Chedorlaomer not only called upon his allies to assist him, but subsequently sent the combined army not just against the Cities of the Plain, but against other nations as well. Wholesale rebellions against an empire often occur in conjunction with a disrupting event, such as the death of a king. It's quite plausible that a king of Elam died and Chedorlaomer became the new king.

King Bera, seeing that the monarch with whom he had made the agreement was no more, likely decided it was the right time to stop paying the hated and crushing tribute.

The new king of Elam would have many reasons to put down a rebellion, not the least of which was the obvious challenge to his rule. If there was reason to doubt the loyalty of the Elamite army, Chedorlaomer would have more than one reason to call upon his allies for support. Once the soldiers of his allies arrived and his rule established, it would be a natural extension to bring the tributary kingdoms back in line and refill the royal treasury. So in the fourteenth year, possibly the fourteenth year of rule of Bera king of Sodom, the allied army comprised of about three or four thousand well armed and armored soldiers, marched south. They likely also brought war chariots, perhaps seen for the first time in the Levant.

Flushed with victory over the Transjordan the allied kings continued their march and next fell upon the Amelikites, possibly a proto-Arabic people who dwelt in the valley city of Kadesh. This city should not be confused with the Syrian city of the same name that became a point of conflict between the Egyptian and Hittite empires a century or so later. This area and Edom mentioned earlier are far south of the Dead Sea and attests to the remarkable march of the Eastern Kings under the leadership of Chedorlaomer.

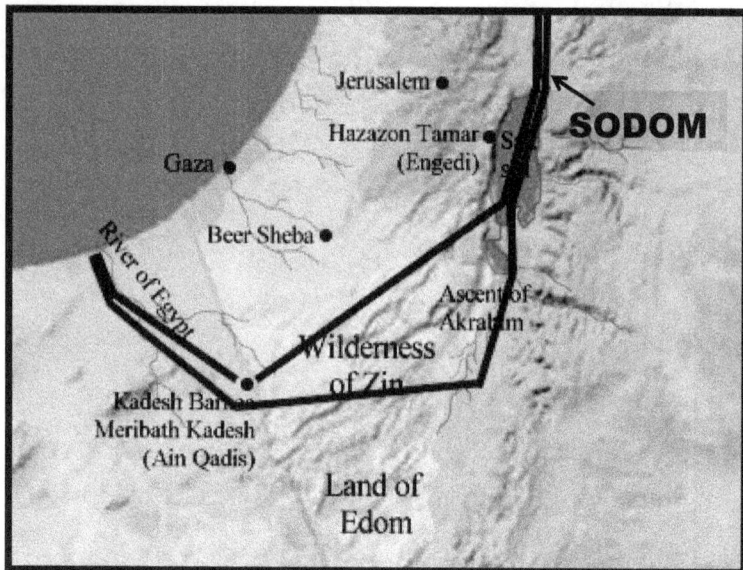

The map above shows the relationship of the Land of Edom as well as Kadesh and Hazezon Tamar from both Sodom and the Dead Sea.

The allied kings then moved back north, striking the Amorites in the town of Hazezon Tamar, also known as En Gedi on the west coast of the Dead Sea. These would be the last people they conquered before moving against Sodom and the other Cities of the Plain.

At this point, we are presented with a curiosity and a question. Having decided to wage war on the east and south side of the Salt Sea as well as part way up the west side, why did the allied kings not continue north along the west coast of the Salt

Sea and attack Hebron and Salem, and Jericho? These three cities would seem to be as attractive as any of the others they had already engaged.

The answer may be that they felt over stretched and too far from their home base to attempt anymore conquest other than what had already been a part of Chedorlaomer's empire. Whatever the reason, they seem to have chosen to march around the south end of the Dead Sea and followed the trade route on the east side of the sea north through Bela (Zoar) to the valley of Siddim.

## Battle of the Nine Kings

According to *Genesis* Sodom and the Cities of the Plain seem not to have been ready for the attack of the allied kings. It's a mystery why they were not. Surely some word of the other city states of the Transjordan falling under Chedorlaomer's assault came to them.

Let's follow along with *Genesis* and see how the events may have played out according to that text.

The allied kings marched through Bela with little to no resistance, for that town's king had already marched north to join King Bera's army. All the fighting men of the Kikkar plain assembled under the banners of Sodom, Gomorrah, Admah, Zeboiim and Bela awaiting the attack of Chedorlaomer. Jericho, the western most city of the plain, does not appear to have responded to the call to arms. And when the allied kings triumphed over the combined soldiery of the Cities of the Plain, Jericho did not face retribution as would Sodom and Gomorrah. Did they purposely hold back? Or did word come to them too late? We will likely never know.

There are no details of the actual battle extant however, we can do some surmising. The mention in *Genesis* of the king of Sodom and the king of Gomorrah falling into slimepits is an indication that king Bera did not deploy his army well. When his army began to disintegrate under Chedor-laomer's slashing chariot attack and heavily armored infantry there was not a clear avenue of retreat. Or if there was an avenue, it was too constricted to handle the onrush of the panicked troops. It's likely that King Bera and King Birsha hid in the pits until escaping back to Sodom, followed closely by the victorious army of the Eastern Kings.

The defeat of the Cities of the Plain was complete. The gates to both Sodom and Gomorrah were thrown open, and the army of Chedorlaomer and his allies took everything, all the goods, all the treasure, and a goodly number of the people. Those taken were likely the young, the strong, and the women. King Bera was left with the old and the frail except those who might have escaped into the wilderness.

Although not specifically mentioned, it is possible that the other cities were also stripped of any valuables or at least had to pay a heavy tribute. However, King Chedorlaomer's army may have been too small to control any more prisoners or haul any more of the spoils. The people taken were loaded with the booty of war and the kings continued their journey. Among the captives were Abram's nephew Lot and his family.

The Roman Jew Flavius Josephus's writing on this subject is more concise and is that of a historian rather than a rabbi:

*AT this time, when the Assyrians had the dominion over Asia, the people of Sodom were in a flourishing condition, both as to riches and the number of their youth. There were five kings that managed the affairs of this country: Ballas, Barsas, Senabar, and Sumobor, with the king of Bela; and each king led on his own troops: and the Assyrians made war upon them; and, dividing their army into four parts, fought against them. Now every part of the army had its own commander; and when the battle was joined, the Assyrians were conquerors, and imposed a tribute on the kings of the Sodomites, who submitted to this slavery twelve years;* [I place the date of this battle circa 1736 BCE] *and so long they continued to pay their tribute: but on the thirteenth year they rebelled, and then the army of the Assyrians came upon them, under their commanders Amraphel, Arioch, Chodorlaomer, and Tidal.*

*These kings had laid waste all Syria* [sic][25], *and overthrown the offspring of the giants. And when they were come over against Sodom, they pitched their camp at the vale called the Slime Pits, for at that time there were pits in that place; but now, upon the destruction of the city of Sodom, that vale became the Lake Asphaltites, as it is called.*[26] *However, concerning this lake we shall speak more presently. Now when the Sodomites joined battle with the Assyrians, and the fight was very obstinate, many of them were killed, and the rest were carried captive; among which captives was Lot, who had come to assist the Sodomites.*

---

[25] This is a reference to the Transjordan city states.

[26] As stated before, the Dead Sea existed thousands of years before the battle of the nine kings. Josephus is merely repeating an erroneous interpretation of *Genesis*.

Although Josephus refers to the invaders as Assyrians, *Genesis* says they were from Elam and its allies which were possibly rivals of the Assyrians at that time. Later, Elam was brought under the Assyrian Empire. It is possible that one or more of King Chedorlaomer's allies was an Assyrian king. It seems, according to *Genesis*, that Chedorlaomer only made one mistake. He took Abram's nephew Lot and his entire family captive.

I have a real problem with this narrative and despite what is written in *Genesis* I think this event played out in a much different manner. What I have titled *The Alternative Battle of the Nine Kings* is one of two possible, and to me, much more plausible accounts than that given in *Genesis*.

## An Alternative Battle circa 1750 BCE

Both the narrative in *Genesis* and by Josephus attest that King Bera and the other kings of the Kikkar were handily defeated. They go on to assert that the allied Army, an army estimated by experts today as numbering 3,000 armed and armored soldiers, then plundered Sodom and Gomorrah. This appears on the surface of it to be somewhat implausible for the following reasons.

The allied army under King Chedorlaomer has marched many hundreds of miles and fought many battles all or most in the heat of the summer. They surely have suffered many, many casualties not only from war, but from exhaustion, heat stroke, and disease. Quite likely by the time they reached the gates of Sodom, that great fortified bastion of a city, they were not in any shape to mount a siege, let alone assault its walls.

It seems more plausible that King Bera and the other Kikkar kings met with the four allied kings and agreed to a settlement. This likely included a resumption of the tribute, perhaps with more gold and silver thrown in as a punitive measure and bolstered by the swearing of fealty to the king of Elam. But if that was the case, how did Lot end up having to go with the invaders?

Likely a part of the terms was the surrendering of a number of captives to be taken back to Elam as hostages and oh, by the way, carry the treasure on the way. It goes almost without saying that King Bera would give up those who only recently had come to Sodom rather than someone with a close and long tie to the city and its people. It was Lot's lot (no pun intended) and that of his family to be among the captives.

*Genesis* only mentions that Sodom and Gomorrah were looted, it doesn't say that Admah, Zeboiim, or even Bela forfeited treasure or goods. This would be in keeping with the weakness of the allied army by the time they arrived at Sodom. So, if Sodom was so strongly fortified, why did they come to terms? Despite being weakened by the long campaign, King Chedorlaomer's army was still capable of inflicting great damage upon the unprotected villages and hamlets around Sodom, not to mention the crops in the fields. King Chedorlaomer likely arrived at Sodom in the early fall, just about harvest time. King Bera and his people could not risk the destruction of their food source for the coming winter. And so King Bera made an agreement that included captivity for Lot. I believe it was this act more than any other that was the *sin of Sodom* and its vilification by the late writers of *Genesis.* Couple that with the fiery destruction of all the Cities of the Plain and you have the basis for *Yahweh's* destruction of them.

The truth is that the terrible destruction of the Jordanic Civilization may have had nothing to do whatsoever with Abram, his people, the wickedness or lack thereof of Sodom, or the god the Hebrews called *Yahweh.* That is a bold statement, and I think I substantiate it as this narrative continues.

## The King of Elam's Raid

Here is the second account I think quite plausible and is even more likely what happened. I have already stated that the army of the allied kings would, at the time they invaded the Land of the Kikkar, been nearly worn out, thinned out, and ready to go home. The combined army of the Cities of the Plain likely matched in numbers or even exceeded the army of King Chedorlaomer. All King Bera had to do was man the walls of his powerfully fortified city and dare Chedorlaomer to do his worst. And I think that is what he did, in effect dealing the allied army its first defeat. The walls of Sodom had served perfectly well in protecting its people for a number of centuries. It made no sense at all for King Bera to risk defeat by deploying his army outside the city. If we assume for the moment that he kept his army inside the city, what was King Chedorlaomer's recourse?

There was only one thing Chedorlaomer could do, lay waste the countryside and gather the up the people who were not inside the city. These people would be residents of the villages and hamlets and the individual farms of the river valley. No doubt numbered among the farmers and herdsmen were the followers of Lot, his animal herds, and his goods. And that is exactly what Chedorlaomer did.[27]

No doubt vowing to King Bera that he would be back during the following year's campaigning season to bring him completely to heel, King Chedorlaomer made off with what his army and their captives could carry. The gates of Sodom and Gomorrah were never thrown open to the enemy. And there is no archeological evidence, at least thus far, that indicates any enemy ever breached or forcibly entered the walls of Sodom.

The walls of Sodom and Gomorrah were close to hand when the allied army arrived from the south. There was the added attraction that they were also near the fords across the Jordan River leading to Jericho. It follows that King Chedorlaomer gathered what people and goods he could from outside the walls of these two cities. And that is why in *Genesis* they are the only cities listed as being looted.

---

[27] *Genesis* supports this idea with the statement *And there came one that had escaped, and told Abram the Hebrew...*this person was without a doubt one of Lot's followers.

During this roundup of captives, one of Lot's people escaped the eye of the enemy soldiers and fled, straight to Hebron and Lot's uncle, Abram.

## The Reaction of Abram

After dividing his flocks and people with Lot, Abram was still the leader of a large tribe of Habiru and had many allies among the local strong men. Abram's reaction according to *Genesis* was swift.

*Genesis 14:11 – 16 And they took all the goods of Sodom and Gomorrah, and all their victuals, and went their way.*

*And they took Lot, Abram's brother's son, who dwelt in Sodom, and his goods, and departed.*

*And there came one that had escaped, and told Abram the Hebrew; for he dwelt in the plain of Memre the Amorite, brother of Eschcol, and brother of Aner: and these were confederate with Abram.*

*And when Abram heard that his brother was taken captive, he armed his trained servants, born in his own house, three hundred and eighteen, and pursued them* [the allied kings] *unto Dan.*[28]

*And he divided himself* [his force] *against them, he and his servants, by night, and smote them, and pursued them unto Hobah,*[29] *which is on the left hand of Damascus.*

*And he brought back all the goods, and also brought again his brother Lot, and his goods, and the women also, and the people.*

It seems pretty clear that Abram indeed had a great victory over the allied kings. It is less likely that he killed the kings or at least not all the kings. What Abram did do was to deal their allied army its second defeat and the first in actual combat. Abram delivered a hard enough blow that he freed his nephew Lot and his family, as well as his followers and herds. In addition Abram recovered the many subjects of King Bera and most if not all the treasure and goods looted by the enemy. It was a major victory over a powerful and feared king.

But beyond these positive events for Lot and Abram lay a much more important political victory for the Cities of the Plain. King Chedorlaomer's hold over the Kikkar valley was broken forever, along with his alliance. Abram's assault may even

---

[28] It is estimated that Abram had about a thousand other warriors from his confederates for a total force of at least 1300 men.

[29] Hobah was located north of Damascus but nothing else is known about it.

have resulted in King Chedorlaomer's death, because after this event he is lost to history. King Bera understood that his own alliance in the Kikkar valley was now poised for even greater freedom and future achievements. But he needed his farmers and herdsmen back, and so he rode to meet with Abram and Melchizedek, the King of Salem, to negotiate for them as told in *Genesis 14:17 - 24.*

*And the king of Sodom went out to meet him after his return from the slaughter of Chedorlaomer, and of the kings that were with him, at the valley of Shaveh, which is the king's dale.*[30]

*And Melchizedek, king of Salem brought forth bread and wine: and he was the priest of the most high God.*

*And blessed him, and said, Blessed be Abram of the most high God, possessor of heaven and earth:*

*And blessed be the most high God, which hath delivered thine enemies into thy hand. And he (Abram) gave him (Melchizedek) tithes of all.*

---

[30] It is thought that this valley lays two and a half miles south of Jerusalem near Beth-Hakkerem so it was likely the valley of the king of Salem.

*And the king of Sodom[31] said unto Abram, Give me the persons, and take the goods to thyself.*

*And Abram said to the king of Sodom, I have lift up mine hand unto the Lord, the most high God, the possessor of heaven and earth.[32]*

*That I will not take from [thee] a thread even to a shoelatchet, and that I will not take any thing that is thine, lest thou shouldest say, I have made Abram rich:*

*Save only that which the young men have eaten, and the portion of the men which went with me, Aner, Eschcol, and Mamre; let them take their portion.*

This was a very bold thing for Abram to do, to more or less spurn a powerful king, a king whose realm his nephew Lot fully intended to return to.

But King Bera had gotten more than he bargained for and in a good way. And so he returned to the Land of the Kikkar with his people including Lot and his family, much better off than he had left. The rejoicing of the entire Kikkar

---

[31] Although *Genesis* refers to the king of Salem by name, it never mentions this king of Sodom by name. It is possible that Bera was killed in the battle with Chedorlaomer and this was a new king of Sodom although I don't think so.

[32] It appears that Abram is declaring to the king of Sodom his allegiance to *Yahweh* so that the Kikkar king knows he has rejected the deities of Sodom and further, he is the sworn man of Melchizedek.

valley most have been long and joyful with feasts and celebrations followed by the determined resumption of their way of life.

Now indeed Sodom and the entire Jordanic Civilization were set to rule in the southern Levant and perhaps even further. Only future events beyond any control of humans in that time and indeed into our own time would shatter the bright future the Lord of the Kikkar envisioned.

But how did Abram do it? How did he defeat an army that had run roughshod over the many different and often powerful, tribes of the Levant?

Before I go into this account I would like to pause a moment and address the possible fate of one of the other allied kings. That king is Amraphel of Shinar. Researchers have asserted this man was actually King Hammurabi of the Babylonian Empire. His listed date of death as 1750 BCE fits well with the time frame of Abram's raid. If Amraphel is Hamurrabi, this was an important event indeed, and here is why.

The name Hammurabi is Akkadian, but this ancient ruler and lawgiver was actually an Amorite and was born with the Amorite name of Ammurapi meaning *paternal kinsman healer*. Hammurabi was the sixth king of Babylon, but more importantly he was the first king of the Babylonian Empire.

At the time the allied army invaded the Levant Elam was a vassal state of Babylon. This suggests that the actual commander of the army was King Hammurabi himself, not King Chedorlaomer. An ancient diplomatic report during Hammurabi's reign demonstrates that he often enlisted the aid of vassal kings.

*...There is no king who is powerful for* [by] *himself: with Hammurabi, 'the man of Babylon,' go 10 or 15 kings...*

There are indications that just before his death in 1750 BCE, Hammurabi was a sick man. Was his invasion of the Levant his last military campaign? It is quite possible. That he died at the hands of Abram's recovery party is debatable, but it would have been a fitting end to the warrior king. His conquests, as impressive as they were, did not long survive his death.

Hammurabi is best known as a lawgiver, and especially what is called *Hammurabi's Code*, one of the first laws written down and placed upon a

The top of this stele depicts Hammurabi receiving his royal insignia from the sun god Shamash. The author has viewed this figure in the Louvre Museum Paris, France.

stone stele in public places for people to read. This ancient king's portrait is often found on and in government buildings to include the U.S. House of Representatives in the nation's capital and on the south wall of the U.S. Supreme Court building.

## The Battle of Dan

When Abram learned of Chedorlaomer's raid on Lot's tribe and the capture of both his nephew and his family he was both alarmed and greatly angered. He quickly sought out his friends consisting of Memre, Eschol and Aner, all Amorites who lived near Abram with their own herds and armed retainers. Abram quickly enlisted their aid and seeing the chance for plunder as well as revenge, they responded by gathering their men. Combined with Abram's more than three hundred warriors the four leaders sat off in pursuit of the Eastern Kings.

King Chedorlaomer and his allies consisting of King Tidal, King Amraphel, and King Arioch were exhausted. Their soldiers were exhausted, their animals were exhausted, their weapons were dulled, wagons and equipment were no doubt in need of repair after the many months of warfare. Despite being refreshed by stolen wine, fresh food, and the comfort of captive women, their soldiers longed for home.

Abram's force was well rested, nourished and highly motivated. They were lightly armed compared to the soldiers and no doubt stripped down to loin cloths and head coverings to move with greater speed. Their arms weighed much less than the soldier's and likely consisted of spears, bows, and long knives and slings. So fast were they that they can even be considered foot cavalry in comparison to the army of the Eastern Kings. The distance covered by King Chedorlaomer's forces in days was traversed by Abram's warriors in a matter of hours.

No doubt the people of Jericho were astonished to see the army from Mesopotamia marching by and even more surprised to see a band of over a thousand warriors stream by later in hot pursuit. The Eastern Kings did not even hesitate or consider attacking Jericho. They were headed home.

Abram moved quickly but stealthily, his scouts keeping visual contact with King Chedorlaomer's rear guard. The enemy was headed north, up the very road that Abram had traversed coming south to the Land of Canaan, and so it was familiar to him. He waited to strike, waiting until the terrain was to his advantage.

When Abram judged the position of the enemy was right where he wanted them, near the site of Dan, he called his captains together. Hunkering down with a stick for drawing he showed them how he wanted to divide their force and attack Chedorlaomer's camp in the early morning hours while darkness still covered the land and the soldiers sleeping in their deepest sleep.

In the pre-dawn darkness with the camp's fires burning low, his stealthiest men crept upon the enemy sentinels, cutting their throats silently and signaling the remainder of the warriors to advance.

Suddenly they were upon them from two directions. Confusion reigned in the camp as the two wings of Abrams' force struck from two different directions, killing soldiers in their blankets and cutting the bonds of captives freeing them to join in the attack.

The disciplined soldiers had no time to join ranks, no time to mount chariots and met their attackers only with weapons that came hastily to hand. It would prove not to be enough.

Abram's men were so quickly upon the army of the Eastern Kings that they caught and killed some if not all of the monarchs. I believe that King Chedorlaomer and possibly King Amraphel of Shinar, who may have actually been King

Hammurabi, were slain. The *Old Testament* leaves no doubt, claiming all the kings were killed and that is certainly possible.

What is certain is that Hammurabi, who called himself King of the Amorites, died about this date and may have been killed in battle by a combined force of Habiru and Amorites. Also certain is that King Chedorlaomer is not heard from again, and no army from Mesopotamia comes seeking a spoil from the Cities of the Plain ever again. Of course, the Cities of the Plain would not last too many more years. However, before their demise they would experience a golden age, a time of growing power and trade, influence in the wider Levant and beyond, and a vast increase in their wealth. It would seem as though the gods were smiling upon them. But the favor of gods can often be fleeting.

# Chapter VI
## The *Genesis* Destruction

Let's take a closer look at the Kikkar Plain and its peoples before the destruction of Sodom and Gomorrah.

The destruction of the Cities of the Plain to include Sodom took place in what historians have dubbed the Middle Bronze Age. This age of man stretched from approximately 2000 BCE to 1550 BCE. Populations contemporary with the Semitic peoples of the Kikkar[33] were roughly as follows: The ancient Egyptians, Assyrians, Akkadians, and the Hittites of the region comprising what is called today the Middle East; the Harappans of the Indian subcontinent, and the Shang Dynasty of China.

The many other peoples mentioned in this history during this time frame in the Levant, were predominantly Semitic in origin. These peoples include but are not limited to the Habiru (Hebrew), Hyksos, Amorites, and a myriad of other Semitic tribes.[34]

---

[33] Kikkar, named for the circular shaped bread of the people refers to the flowing area of land, round in its general outline that encompasses the valley of the Jordan River.

[34] This does not include the Philistines who were likely of Greek origin.

This is the *Genesis* account of the destruction of Sodom and Gomorrah.

*18:20 And the Lord said, Because the cry of Sodom and Gomorrah is great*[35]*, and because their sin is very grievous;...*

*19:24 Then the Lord rained upon Sodom and upon Gomorrah brimstone and fire from the Lord out of heaven;*

*19:25 And he overthrew those cities, and all the plain, and all the inhabitants of the cities, and that which grew upon the ground.*

Abram (later Abraham) saw this happen:

*19:28 And he looked toward Sodom and Gomorrah, and toward all the land of the plain, and beheld, and lo, the smoke of the country went up as the smoke of a furnace.*

That is the explanation for the death of thousands of people and the complete devastation of a land known as the Kikkar Plain. The Kikkar

---

[35] Rabbinical writings centuries after the destruction of Sodom (1552 CE) claim this *cry* was raised by a daughter of Lot named Peleith or Paltith who was burned to death by the people of Sodom for giving food to a starving man. There is no mention of this event in *Genesis* indeed it doesn't assign a name to any of Lot's daughters.

is a plain in the valley of the Jordan River primarily located today in the *Hashemite Kingdom of Jordan*. The word *kikkar* is a reference to the shape of bread made by the people who lived on the plain, it also means *circular in shape* and when used to describe a geographical area means a spreading section of land in the general shape of a circle. And that is the description of the Jordan River valley to this day.

This was a most desirable area and its virtues are expounded upon in Genesis:

*13:10 – 11 And Lot lifted up his eyes, and beheld all the plain of the Jordan, that it was well watered every where, before the Lord destroyed Sodom and Gomorrah, even as the garden of the Lord [Eden], like the land of Egypt...*

The last sentence *like the land of Egypt...* is comparing the well watered delta of the Jordan to that of the Nile River, life blood of the civilization of ancient Egypt. This was truly a desirable place to live, for in addition to the waters of the Jordan many springs of fresh water sprang up in the valley from the watershed created by the nearby mountains. Indeed other, smaller rivers flowed around Sodom and near the other cities of the

Kikkar. These cities came to support a large population and gave rise to a culture unique to the area that I have chosen to call the Jordanic Civilization.[36]

From my earliest readings of the fate of Sodom as a young man I knew that what *Genesis* was describing was nothing less than an explosion of unthinkable magnitude, especially for the 18th century BCE. Indeed the first thought that came to my mind was the similar destruction of Hiroshima and Nagasaki by the atomic bombs dubbed *Fat Man* and *Little Boy*.

Later, after reading about a number of events across time and the globe involving the air burst of asteroids and their impact upon our world I began to speculate this could be the explanation for the demise of an entire civilization.

Now, through the efforts of archeologist Dr. Steven Collins, his colleagues and many volunteers, it seems my thoughts in this regard are fast approaching confirmation. I very much recommend that if you desire to learn even more about the Jordanic civilization in the Kikkar, that you read Dr. Collins' book *Discovering the City of*

---

[36] To my knowledge I coined the phrase "Jordanic civilization".

*Sodom* published in 2013 by Howard Books, of Simon & Shuster Incorporated.

## The Sins of Sodom and Gomorrah

Jewish rabbinic traditions hold that the sins of Sodom and her daughter cities were sins of property. They state that the inhabitants of the cities all believed that "what's mine is mine, and what's yours is yours". This is in keeping with the views of most modern states and peoples today. Yet the Hebrews found this to be a blatant lack of compassion and a sin worthy of mortal punishment by *Yahweh*.

I need to point out here that at the time of its destruction, Sodom and indeed the entire region of the Ancient Near East (ANE) had just experienced a devastating and long lasting drought. Even *Genesis* makes note of this as it regards Abram's decision to by-pass the *Promised Land* after his arrival near Bethel, in favor of Egypt because of this selfsame drought:

*Genesis 12:9 And Abram journeyed, going on still toward the south.*

*12:10 And there was a famine in the land: and Abram went down into Egypt to sojourn there; for the famine was grievous in the land.*

In light of this vast and devastating drought, the attitude of the Kikkar inhabitants seems to be justified. They no doubt were having problems surviving themselves, let alone anyone else that happened to come visiting. Other sins outlined for the justification of Sodom's destruction seem more, for lack of a better word, ridiculous based on these stories from rabbinical writings.

A servant of Abram named Eliezer[37] traveled to Sodom to visit Abram's nephew Lot. While there he got into an altercation with a citizen of Sodom who, subsequently in his rage, struck Eliezer with a stone in the forehead. When Eliezer took the matter to a Sodom judge, the judge sided with the man of Sodom and ordered the Hebrew to pay the man for a medicinal bloodletting. This false judgment in the eyes of Eliezer caused him to become so angry that he threw a stone and struck the judge on his forehead for which the servant of Abram demanded the judge make the payment to the man of Sodom. If he had done this act, Eliezer

---

[37] The name Eliezer means "God is help" and he was purportedly the steward of Abram's household.

would likely have at least been imprisoned for assaulting a town elder.

Another rabbinical sin attributed to the people of Sodom was the giving of marked money to poor people. Merchants, seeing the secret mark would not accept the money for food or goods, causing the bearer to starve to death. The obvious question would seem to be why didn't the individual that was denied goods or services simply leave Sodom and go elsewhere to spend the money, perhaps to one of the other nearby cities?

In a similar vein is the story of the Sodom bed, to which guests to the city were led and when they lay down they were measured by the bed. If too short they would be physically stretched until they fit. If too long they would have their legs cut down until they fit. Presumably if, as in the story of Goldilocks and the three bears, if you were "just right", you were left to sleep comfortably.

Both stories are implausible and seem hardly to attain the definition of high crimes seemingly needed to justify the destruction of your entire civilization. But then, there is a final story.

In this story a daughter of Lot, presumably a third daughter other than the two who would later have incestuous sex with their father according to *Genesis*, provided food and comfort to a beggar.

For this act she was condemned and burned at the stake. It is her cries *Yahweh* purportedly hears in this passage taken from *Genesis*:

*Genesis 18:20 And the Lord said, Because the cry of Sodom and Gomorrah is great, and because their sin is very grievous;*

*18:21 I will go down, and see whether they have done altogether according to the cry of it, which is come unto me; and if not, I will know.*

If I were the prosecuting district attorney in this case for the condemnation of an entire civilization and the execution of thousands upon thousands of people, I would refuse to prosecute.

Someone will say, but what about the passage in *Genesis* where the men of the city wanted to have sex with the angels *Yahweh* sent to investigate and possibly destroy them? Okay, let's take a look at the passage.

*Genesis 19:4 But before they* [the angels] *lay down* [in the house of Lot] *the men of the city, even the men of Sodom, compassed the house round, both old and young, all the people from every quarter:*

*19:5 And they called unto Lot, and said unto him, Where are the men which came in to thee this night? bring them out unto us, that we may know them.*

This does not sound like a demand to have sex with the angels, it sounds like a demand from the city's night watch and *"...all the people from every quarter."* to know who has come into their city. In fact, it is Lot who seems to bring up sex when he says:

*Genesis 19-8 Behold now, I have two daughters which have not known man; let me, I pray you, bring them out unto you, and do ye to them as is good in your eyes; only unto these men do nothing; for therefore came they under the shadow of my roof.*

This sounds like a rather crude attempt to bribe the city guard to secure and hide the agents sent to destroy the city. Again I see nothing in the actions of the citizens of Sodom which includes not only young and old men but women as well, which warrants the destruction of a single city, let alone a half dozen or even more.

Here is how the Jewish – Roman historian Flavius Josephus renders the account of Sodom's sins and the wrath of *Yahweh*.

*About this time the Sodomites grew proud, on account of their riches and great wealth: they became unjust toward men, and impious toward God, insomuch that they did not call to mind the advantages they received from him: they hated strangers, and abused themselves with Sodomitical (sic) practices. God was therefore much displeased at them for their pride, and to overthrow their city, and to lay waste their country, until there should neither plant nor fruit grow out of it.*

Again, I must say I fail to see any justification for *Yahweh's* supposed destruction of Sodom.

What I do see here is an attempt to explain a rare but natural event of cosmic origins, by claiming the destruction to be at the hands of *Yahweh*. What I do see here is the work of religious men who wanted to instill obedience to *Yahweh*, and by extension to their, authority. You can't have a functioning priest system unless you have obedience of the flock, and their willingness to support the priesthood through their donations to the temple and therefore by extension, to *Yahweh*. It is even possible, indeed probable that the attribution of Sodom's destruction to *Yahweh* propelled this former Hyksos Storm God to the vaulted pinnacle he would attain from that time forward.

The story of the destruction of Sodom and the rest of the Jordanic civilization is real. But the explanation given in *Genesis* as to why it happened is suspect. However, the description of the event itself is and remains the only account of a city or group of cities destroyed by a cosmic event originating from the heavens. As such, it is an event whose aftermath both immediate and long term, should be studied by Earth's governments today. It may well foretell our own future.

# Chapter VII
## The Cosmic Destruction

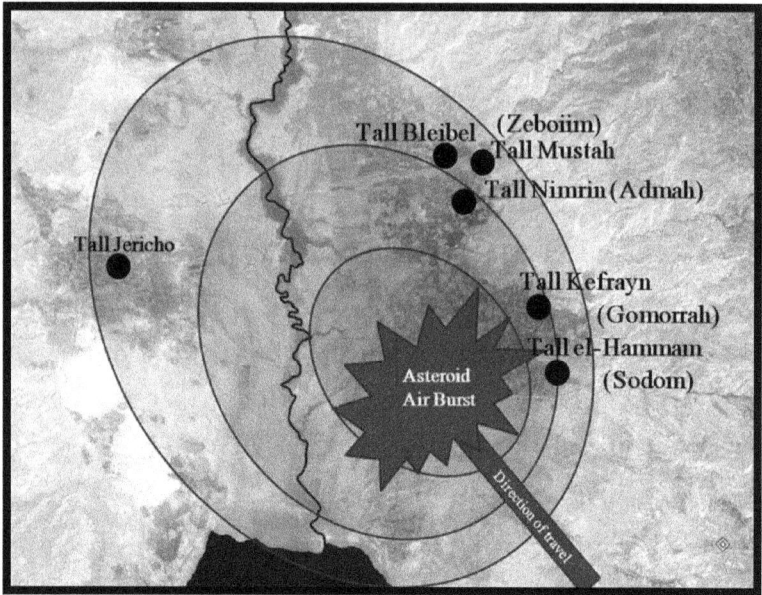

Map of the likely asteroid impact zone. This map is adapted from a similar map created by Dr. Collins staff. I believe this modification better fits the evidence of a massive destruction event that included Jericho.

The unofficial name of this asteroid is Sodom1760BCE.[38] The destruction date is about the year 1700 BCE in and during the Middle Bronze Age 2 period.

---

[38] Conditionally assigned by the author.

The projected ground effects are of an airburst chondrite asteroid traveling at 40,000+ miles per hour, delivering 50 kilotons[39] of TNT detonation at an altitude of 2,200 feet, on an early summer morning in the year 1700 BCE that includes a fireball, shock wave, and air blast.

This event was immediately lethal for all life forms including humans and large animals within the inner six kilometer wide ring. This includes the cities of Sodom and Gomorrah. All buildings are crushed down and completely incinerated and most of the population vaporized where they stand.

Severe lung damage occurred inside the next 10 to 12 kilometer circle due to pressure pulse (concussion). Any and all humans or animals are unlikely to survive, with universal injuries and fatalities widespread.

The air blast demolishes even heavily built and fortified stone buildings out past the second ring to a distance of 7.97 miles to include the cities of Admah, Zeboiim, and Jericho.

Residential buildings collapse and wooden houses, exposed beams etc. are ignited by thermal radiation to a distance of 14+ miles. Third degree

---

[39] A recent scientist during a presentation on a possible asteroid impact on Sodom estimated the detonation may have released up to 200 Kilotons of explosive force. If his estimate is close, the destruction would be much more widespread than I have outlined.

burns to exposed human flesh is a 100% probability with mortality immediately or soon following after.

All vegetation is ignited, burns furiously, and is completely consumed out to the end of the outermost 50 kilometer ring. Trees directly under the air blast are stripped of leaves and limbs and ignited and those further out are blown down in all directions and ignited. Those humans and animals not killed outright have sustained life threatening injuries and without immediate and organized medical attention, which is not available, will succumb to their wounds. The Jordan Valley is part of the Rift Valley[40] and a massive earthquake caused by the asteroid's detonation rolls through the valley and up the walls of the surrounding ridges, toppling any buildings left standing.

The Jordanic Civilization of the Jordan River valley (the Kikkar) is destroyed. The number of deaths are conservatively estimated in the tens of thousands as follows: In and around Sodom 50K, Gomorrah 20K, Admah, 25K, Zeboiim (combined two towns) 20K, and Jericho 20K for a rough estimated total of 135,000 souls that perished in

---

[40] The Rift Valley is a geologic depression of southwest Asia and eastern Africa extending from the Jordan River valley to Mozambique. The region is marked by a series of faults caused by volcanic action.

this cosmic event.[41]   That is a large number of people lost at once for this time period, perhaps the most ever until perhaps the time of the Crusades.

Cities and towns further away such as Salem, and Bela, Bethel, and possibly Ai (if occupied), experience sound effects louder than any ever produced by lightning in their experience.  Then a shock wave similar to an earthquake that causes some walls to topple and a wind hotter than that from the southern deserts and powerful enough to blow many animal skin tents down and ignite exposed wooden objects as well as inflict burns on exposed flesh.

Visual effects of the detonation are visible before and after detonation.  This includes the thunderous crash of the asteroid breaking the sound barrier followed by the furious roar of its passage through the air and the sound of the airburst detonation over the Kikkar plain.  These sound effects alone are likely to have produced hundreds and perhaps even thousands of humans traumatized by fear and prostrated with disabling psychic shock.  The rising of the fiery clouds, dense black smoke and tons of debris into the air would likely drive them to their knees in

---

[41] Admittedly this is only an estimate, but seems reasonable, at least at this time, pending further excavation of the Kikkar plain.

supplication to whatever deity they worshipped. The roiling black clouds were rife with tremendous lightning bolts follow by crashing thunders. The following passage is from *Genesis* Chapter 19: 24, 25, 27 and 28:

*Then the Lord rained upon Sodom and upon Gomorrah brimstone and fire from the Lord out of heaven;*
*And he overthrew those cities, and all the plain, and all the inhabitants of the cities, and that which grew upon the ground.*
*And Abraham gat up early in the morning to the place where he stood before the Lord.*
*And he looked toward Sodom and Gomorrah, and toward all the land of the plain, and beheld, and, lo, the smoke of the country went up as the smoke of a furnace.*

Author Annalee Newitz wrote a book whose title alone tells what to do in a case such as the destruction of the Jordanic Civilian: *Scatter, Adapt, and Remember, How Humans will Survive a Mass Extinction.*

And that is exactly what Abram proceeded to do. He did not look for nor wait for his nephew

Lot and his family who lived in Sodom, for he assumed them to be dead. This event literally put the fear of *Yahweh* in Abram and he fled. He fled from the *Promised Land* to the southern desert and to the country known as Gerar. He took all his people with him and I have no doubt that many other people fled from the terrifying events they witnessed. *Genesis* Chapter 20 reads:

*And Abraham journeyed from thence toward the south country, and dwelled between Kadesh and Shur, and sojourned in Gerar.*

The destruction of Sodom and the cities of the plain was a life changing event for Abram. How serious it was for him is demonstrated by the fact he spent the remainder of his life away from the *Promised Land* proper and sojourned well to the south.

The destruction of the Jordanic Civilization, attributed to *Yahweh* in the manner described recalls the words from the Hindu scripture from the Bhagavad Gita[42]:

*If the radiance of a thousand suns*
*Were to burst at once into the sky,*

---

[42] A 700-verse scripture that is part of the Hindu epic Mahabharata.

*That would be like the splendor of*
*the Mighty One...*
*Now, I am become Death, the*
*destroyer of worlds.*

Dr. Robert Oppenheimer, sometimes referred to as the "Father of the Atomic Bomb" recited this passage upon witnessing the detonation of the first atomic bomb at the Trinity detonation site in New Mexico.

# Chapter VIII
## The Aftermath of Destruction

### Abram, Lot and Lot's Wife

Most people are familiar with the story of Lot just before the destruction of Sodom and the other cities. As the story is related in *Genesis* the angels sent by *Yahweh* warned Lot of the immediate strike against the city where he and his family had dwelt for many years. They told him to flee to the mountains to save himself and his family. But Lot pleaded for the sparing of the city of Bela, also known as Zoar, in which he could find refuge for his wife, daughters, and himself. The angels relented and promised this small city would be spared for Lot's sake.

And so they fled, but Lot's wife, against the express warning of the angels, looked back at Sodom, perhaps in regret for leaving her home, and witnessed the city's fiery demise. For this very human act of showing her regret, she was turned to salt. Let's pause here a moment and consider this.

There is nothing in the explosion of an asteroid that can turn a human or any other living being into a pillar of salt.[43]   If we consider this a valid event there are possibly two things that this passage is referring to and that is either fear, or ash.

The things she saw happening before her eyes brought her up short with a mind numbing, paralyzing fear that brought her to a virtual standstill.  Like the scientists centuries later that witnessed the first atomic bomb detonation, she was innocent and ignorant of what she saw coming at her.

What she saw was a ground rippling shock wave followed close behind by a towering wall of dust and debris with lighting forking through it all running ahead of a roiling red and yellow maelstrom of fire.  In its terrible beauty and deadly majesty it stopped her in her tracks like she'd become stone or a pillar of salt.  Of course she was physically destroyed in an instant of time, she did not suffer other than the mental trauma she had already experienced.

There is the possibility that if Lot's wife were directly below the initial fireball, she could have

---

[43] Lake Natron in Tanzania has so much salt and soda in it that its waters cause any creature such as a bat or bird that falls into it to calcify and is perfectly preserved as they dry.  This process is recorded in the fantastic work of author/photographer Nick Brandt in his book "Across the Ravaged Land".

been burned completely to ash, so quickly she remained standing as a pillar of gray/black ash for a time. Yet, if Lot and or his daughters were close enough to witness this, they would not have survived either.

In the meantime, Lot was herding his two daughters along toward shelter and perhaps some protection from the wrath of the cosmic event unfolding behind them.

I dismiss this story in its entirety as apocryphal. I am convinced that Lot, his wife, and their two daughters died with the thousands of other inhabitants of the Kikkar plain. We need to return to the abduction of Lot and his family by King Chedorlaomer to see why I believe that.

You will recall that when Abram learned of Lot's capture he immediately gathered his own warriors and enlisted the help of his allies and their fighting men to pursue and rescue Lot. He did not hesitate to weigh the consequences of what might happen when they clashed with the trained professional army of the Eastern Kings, he sped ahead. He left everything behind with the single-minded purpose of rescuing his only kinsman. Lot was the son Sara had never been able to give him. He was not going to let him die in slavery.

Likewise, after the cataclysmic events in the Kikkar valley if Abram had thought, suspected, or even heard a rumor that Lot had survived, he would have mounted a search for him. He would not have left a stone unturned in looking for the man the Bible often referred to as Abram's brother! He would not have left so beloved a member of his family behind, unless of course he was convinced of his death.

But Abram did not look for Lot or his family because he knew in his heart and intelligence brought to him that they had not survived. Instead, with the fear of *Yahweh* in him, so to speak, Abram fled Hebron for the southern end of the Levant. Fleeing to the land of the king of Gerar, Abram never returned to Hebron until it was time for him to die. In his book *A Forgotten Kingdom* Leonard Woolley states that when Abram fled to Gerar it was in fact a moral indictment of *Yahweh's* destruction of Sodom and the other Cities of the Plain. And I agree with that determination.

Indeed, Abram was seemingly undergoing a crisis in his faith. Either not trusting *Yahweh* to provide for him and his people, or out of fear of him, he returned to his old ways. Upon reaching Gerar Abram perpetrated the same deception on

King Abimelech that he had on Egypt's pharaoh. He allowed Sarah to be viewed publicly and announced she was his sister. King Abimelech, did lust for Sara because of her exceeding beauty. He took Sara, but soon returned her upon being informed she was Abram's wife. Abimelech had nearly committed the mortal sin of adultery and to make amends gave Abram many valuable gifts including precious metals.

There is only one explanation for Abram's action. Lot, and his entire family, died in the destruction of their adopted home city of Sodom.

However, if that is true, why the elaborate although unbelievable story? Now, we get into the realm of poetic license taken by the author of this part of *Genesis* and his desire to show that Sodom and Gomorrah was indeed destroyed by *Yahweh* and that *Yahweh* was merciful to those who served him. In this case Abram's kinsman Lot.

As for the pillar of salt story, that was to reinforce the idea that *Yahweh* was to be obeyed, whether his word was delivered by an angel or a rabbi. This story is a parable told and retold and finally written down a thousand years after the cosmic event. The tradition of telling parables to illustrate a point is centuries old among the

Hebrew and other peoples to include the greatest parable teller, Jesus of Nazareth.

Hebron is but a short distance of approximately 60 kilometers from Sodom and *Genesis* clearly states that Abram saw what happened in the Kikkar valley.

Abram stayed so long in the land of Gerar that it was there that *Yahweh* made his covenant with him and renamed him Abraham.

# Chapter IX
## The Cosmic Destruction of the Transjordan

Archeologist Nelson Glueck in the Holy Land

During the summer of 1939 archeologist Nelson Glueck surveyed the region east of both the Jordan River and the Kikkar Plain known as the Transjordan. The area he looked at is the very location that King Chedorlaomer conquered piecemeal from Ashtaroth and Ham in the north to El-paran[44] in the south. Chedorlaomer and his allied kings who Glueck refers to as the Eastern Kings, captured all the fortified cities of this desert realm as already outlined, meeting no concerted opposition.

Visiting all the sites of these former small city states he could find, Glueck was amazed at the near total destruction he found. He wrote:

*...such a thoroughgoing destruction visited upon all the great fortresses and settlements of the land...that the particular civilization they represented never again recovered. The blow it received was so crushing as to be utterly destructive. Its cities were never rebuilt... Permanent villages and fortresses were no longer to rise upon the face of the earth in this region till the beginning of the Iron Age.[45]*

---

[44] Modern day Mecca.
[45] Glueck, Nelson *The Other Side of the Jordan.* 114

Glueck of course ascribed this destruction to the Eastern Kings under the leadership of Chedorlaomer. Reading his interpretation today with the knowledge of the discovery of Sodom and its fiery demise, the actual reason for this vast and complete destruction becomes clear.

Many, if not all of these cities, were destroyed when Sodom was destroyed and from the same cause. This is an important finding, because it not only relocates the probable epicenter of the detonation, but also increases its likely magnitude.

This information suggests that the airburst of the asteroid likely occurred above and slightly east the Transjordan ridge. The resulting effects of the blast was felt west to Jericho and beyond and east to the city states of the Transjordan plain. This speaks to an even more massive asteroid than Dr. Collins or I have supposed.

Glueck places the date of destruction of the cities of the east at + or - 1900 BCE. That places this event within two hundred years of my own destruction estimate time for the Cities of the Plain.[46] Glueck's next words, written in 1939, reflect those of Dr. Collins in 2013. But Glueck

---

[46] 1700 BCE.

was also flummoxed by the level of destruction and wrote:

*What prevented the entire land of Transjordan from being settled at least...between the end of the 20th and beginning of the 13th century BC [sic] ...?*

The answer is of course, the same thing that kept the Cities of the Plain, except Jericho, from resettlement during the same 700 hundred year time frame, fear and superstition.

The survivors of the cosmic event fled the area and who can blame them? They had suffered under the conquering army of the Eastern Kings their people enslaved, their fortified sites dismantled or burned, their crops either confiscated or burned and their soldiers killed. When the remaining people returned to the fields and rebuilt their cities they had but a short reprieve before the assault from the very heavens themselves. They left, leaving the devastated lands to tent dwelling, nomadic Bedouin tribesmen. No cities would rise again for at least six hundred years and even longer. Dr. Collins believes that the asteroid blast effected between 500 and 800 square miles. If *Yahweh* wreaked his wrath upon Sodom, there was great collateral

damage to those people and cities around the
intended target.

Probable Asteroid airburst location based on Glueck's
Transjordan archeology.

An explosion above the ridge as opposed to just
above Sodom would certainly send a wave of
destruction and fire well into the eastern
Transjordan. That such a wave could destroy
many of that area's cities and fortresses seems
somewhat obvious and would be well within the
500 to 800 square mile radius of Dr. Collins'
estimate. This was truly a much more massive

event than the mere obliteration of Sodom and Gomorrah. This is yet another indictment of the biblical account of what happened in the Levant of four thousand years ago.

Once again an exceptionally hot fireball smashed into the dwellings of the people on the Transjordan plain. It burned all crops, killed all large animals and people in the open either outright or eventually from searing burns. The shockwave followed in a split second beat of time and blasted the mortar-less stones of their buildings, monuments, and fortresses apart and scattered them like Lego blocks. Bedouins tenting on the plain much like Abram's Habiru also perished, their animal skinned tents becoming their funeral pyres. Anyone that survived fled the area, never to return in their lifetime and not even in several lifetimes. It was truly a widespread and devastating thermal catastrophe that changed the destiny of an entire land and people.

Some may have seen the asteroid approaching arrayed in fire as bright as the sun and bringing the brimstone of yellow fire. Those further away would have felt the concussion, heard the detonation and witnessed a sight all too familiar for those of us in the twenty-first century CE, a mushroom cloud!

# Chapter X
## The Excavation of Sodom

*...Extensive research, along with archeological data from four seasons of excavation, are now leading many scholars to entertain or adopt this theory* [that Tall el-Hammam is biblical Sodom] *on its evidential merits. That the enduring and powerful presence of Tall el-Hammam and its associated towns and villages of the eastern Jordan Disk* [the Kikkar] *during the Bronze Age gave rise to the Cities of the Plain tradition reflected in the stories of Genesis 10-19...*[47]

The Tall el-Hammam Excavation Project is a joint archeological dig agreement between Trinity Southwest University in Albuquerque, New Mexico and the Hashemite Kingdom of Jordon.

The dating method, or chronology applied to the dig was modified and is as outlined earlier. The Chalcolithic Period is a phase of the Bronze Age in which the addition of tin to copper to form bronze during smelting was yet unknown by the somewhat primitive metal workers of the time.

---

[47] Page 24, Season Four Excavation Report, Tall el-Hammam Project, 2009.

The very first season of excavation lasted less than a month, but human remains were found in the MBA (Middle Bronze Age) layers of the tall and left in place for later study. Outlines of the ancient city were established. Exploratory digging demonstrated there were no Late Bronze Age artifacts, indicating there was no immediate reoccupation of the site after its destruction for several hundred years. The Sodom of *Genesis* was, with little doubt, a Middle Bronze Age 2 city. Artifacts found were from the Early Bronze Age Intermediate Bronze Age, Middle Bronze Age, and then skipped over the Late Bronze Age to the Iron Age. Preliminary excavations at the other suspected Cities of the Plain also show this skip over the Late Bronze Age.

One of the more interesting items found was identified as a scarab from the Hyksos, those Canaanite peoples fated to invade and conquer the Nile Delta area of Egypt.

An Egyptian depiction of Canaanites who later became known to the Egyptians as the Hyksos or *foreign rulers*.

The Hyksos first appeared in Egypt as invaders circa 1800 BCE to become the rulers of Lower Egypt. Although the Hyksos invasion would cost Egypt half its kingdom for more than two hundred years, it also brought them improvements. Besides introducing the chariot and the powerful horse, the Hyksos brought new bronze working techniques, pottery making processes, new animals and new crops. They also brought the worship of their Storm God, *Yahweh*, and they fought and conquered in his name. Among the diverse people making up the Hyksos was a tribe diverse in their own right and known as the Habiru or Hebrew.

But the introduction of the chariot would directly affect the Egyptian military and would refine its design and make it a superb fighting platform. So successful would they become in handling the chariot it would allow them to greatly expand and build their greatest empire under Thutmose III.

Sodom existed through the Early Bronze Age, the Intermediate Bronze Age, and the Middle Bronze Age as an organized city-state of a significant size and importance, wielding much influence in southern Levant. It was a large, well-fortified city with an appropriate sized population. It was the leading metropolis of the Cities of the Plain.

The human remains found in earlier excavations were removed for study and they tell an interesting story. Consisting of two adults and one child, Dr. Steven Collins, regarding these remains wrote:

*...What we see are bodies wrenched around in a facedown position, as if they were thrown down in the process of turning away from something—in an unconscious reaction, as if protecting themselves.*[48]

---

[48] *Discovering the City of Sodom* p. 179.

The conclusion drawn was that these people died in *extreme trauma* (Collins) with the charring of bones quite evident. The surroundings of these remains had lain undisturbed for the better part of four thousand years!

Sodom, consistent with the remains of the other Cities of the Plain, lacks any Late Bronze Age architecture and even pottery shards from the LBA are rare. This reinforces the apparent demise of the Middle Bronze Age incarnation of Sodom and her attendant cities of Gomorrah, Admah, Zeboiim, and Jericho. This suggests that all of the Cities of the Plain ceased to exist at the same time except Jericho.

The fortified city-state of Middle Bronze Age Sodom was a huge, complex, and doubtless impressive, capital of the Kikkar. When it ceased to exist after the violent cosmic conflagration that consumed it, the loss impacted not just the Cities of the Plain, but brought death and destruction to the entire region of the southern Levant.

Basalt statue of Thutmose III he became known to scholars as the Napoleon of Egypt, conducting many campaigns of conquest that built the Egyptian empire.

The psychological trauma for those witnessing and perhaps surviving the terminal event is well demonstrated by the fleeing of Abram and his followers with no apparent attempt to discover the fate of Lot and his family. His fear of *Yahweh* and what this almighty deity had done, according to priests such as Melchizedek, caused Abram's precipitous flight to Gerar never to return until he neared the time of his death.

It was no doubt clear to Abram that Lot had perished along with the other citizens of Sodom. His resurrection in the story of his flight to Bela (Zoar) and the begetting of children upon his daughters is a fiction perpetrated later against the supposed incestuous bastardization of the Moabites and Ammonites who resisted Israel's incursion into their land.[49]

---

[49] Genesis 19:30 And Lot went up out of Zoar, and dwelt in the mountain, and his two daughters with him; for he feared to dwell in Zoar: and he dwelt in a cave, he and his two daughters.
19:31 And the firstborn said unto the younger, Our father *is* old, and *there is* not a man in the earth to come in unto us after the manner of all the earth:
19:32 Come, let us make our father drink wine, and we will lie with him, that we may preserve seed of our father.
19:33 And they made their father drink wine that night: and the firstborn went in, and lay with her father; and he perceived not when she lay down, nor when she arose.
19:34 And it came to pass on the morrow, that the firstborn said unto the younger, Behold, I lay yesternight with my father: let us make him drink wine this night also; and go thou in, *and* lie with him, that we may preserve seed of our father.

This fits with the fact that the *Old Testament* is based upon history shaped by ethnic, social, and cultural (primarily religious) considerations in the interpretation of both human and natural events. I have never, and do not now, accept the writings of the *Bible* literally.

Dr. Collins' most recent excavations (2013) have uncovered what he describes as artistic motifs and architectural features that suggest a connection to the Minoan civilization.[50] This is an important historical connection that demonstrates the cultural reach of both Minoa and the Jordanic civilization.

The first fortification of Sodom was a mud-brick city wall built around the lower city that was more than five meters (approximately 16 feet) in thickness. Pottery sherds in the brick mix and foundation allow for an accurate date of construction. These sherds indicate a construction date circa 3000 BCE during the Early Bronze Age 2 (EB2). The wall contained gates, guard towers,

---

19:35 And they made their father drink wine that night also: and the younger arose, and lay with him; and he perceived not when she lay down, nor when she arose.

19:36 Thus were both the daughters of Lot with child by their father.

50 The Bronze Age Minoan civilization arose on the island of Crete in the Mediterranean Sea and was extant before and after the time of Sodom's destruction. The Semitic place name of Kaftor or Caphtor are references to Crete.

an exterior roadway, and market/religious plazas just inside the main city gates.

Approximately 300 hundred years later, around the start of EB3 (circa 2700 BCE) a new more massive wall was built of a solid stone foundation and extending upward in excess of thirty-three feet! It is likely the city wall was built with a rampart of gaps or crenellations that would allow the shooting of arrows, casting of spears, or dropping of stones on attackers, yet provide some protection for city defenders. This was one sturdy wall that proved its worth for the next nine hundred years before new defensive fortifications became necessary.

Despite the city's obvious strength, according to *Genesis* it would fall to an enemy army not by invasion over its walls, but because Sodom's army met defeat outside the city walls. This left the magnificent walls virtually undefended when Chedorlaomer's victorious allied army marched upon it.

But in the end, all of Sodom's formidable defenses would mean nothing when death came to the city from a totally unexpected direction, from out of the very heavens themselves! How could and did this happen?

## Creation of the Angel of Death

The asteroids formed in a belt or orbit between Mars and Jupiter more than four billion years in the past. Unlike the planets of the Solar System, the asteroids failed to come together and only managed to create four large planetoids. These are Ceres, Vesta, Pallas, and Hugiea which together contain about fifty percent of the mass of the asteroid belt, with Ceres (950 kilometers in diameter) designated a dwarf-planet. The rest of asteroids are smaller and range in size down to a dust particle. Asteroids of the belt are constantly moved towards Earth through orbital conflicts with Jupiter, providing new near-Earth asteroids that can pose a threat.

Other asteroids include the Trojans, a minor planet that shares an orbit with a planet or moon, but does not collide with it because it orbits around one or the other of the two *Trojan* points of stability. Some planets have dozens or even hundreds of Trojans. The asteroid 2010 TK is confirmed as the only known Earth Trojan asteroid (confirmed in 2011) and follows ahead of Earth in its orbit around the sun.

Another group of asteroids are the near-Earth asteroids. These are defined as objects in an orbit near Earth that lack the tail or coma of a comet. As late as June 2013, nearly 10,000 near-Earth asteroids are known. These range in size from one meter to thirty-two kilometers (approximately 20 miles) wide. Astronomers estimate that the near-Earth asteroids that exceed one kilometer in width are almost 1,000 in number. The composition of near-Earth asteroids mirrors those of the asteroid belt. There are a few near-Earth asteroids that are burned out comets. Although an asteroid swinging out of the asteroid belt can cause a threat, by far the biggest threat comes from these near-Earth asteroids.

Among the more than 1,000 near-Earth asteroids are recorded, at this time, 160 asteroids that, upon entering the Earth's atmosphere, could cause a global disaster. One of these sizes of asteroid struck the Earth 65 million years ago.

The resulting earthquake was more than 1,000 times more powerful than anything we have ever recorded. It threw down mountains, uplifted seas, and drained lakes. The ejecta (debris thrown up from the surface such as rocks, boulders etc.) rose as high as fifty miles and attained speeds fast enough to escape Earth's gravity and entered outer space. Some may even have traveled to the moon and impacted there. The debris that didn't escape from Earth's pull fell back, on fire from the friction of passing through the atmosphere, and ignited fires around the globe. Sulfur dioxide interacting with nitrogen oxide and water molecules created acid rain that injured and destroyed plants and aquatic animals. However, burrowing animals and those underwater survived better than animals that dwell upon the surface. Up to 70% of all species were killed at the time of impact, most notably the dinosaurs, and in time the percentage rose to more than 90%. This mass extinction however, paved the way for the evolution of our own species.

Ages in the past, even perhaps millions of years ago, an asteroid detached itself from its companions in the Mars-Jupiter belt and swung in a long graceful arc toward the Sun and the Earth. It may have settled into an unstable solar orbit that eventually lined it up, like the sights on a rifle, and fired it toward the third planet from the Sun. It was this possibly rock-iron-nickel composed chunk of mass that entered the atmosphere sometime around 1700 BCE. Its chance encounter with the blue marble planet would also bring it into a chance encounter with a city on Earth, a city the Hebrews called Sodom. But it destroyed much more than one city as it struck. The impact of the asteroid also erased Gomorrah, Admah, Zeboiim, and most of Jericho in the Kikkar Plain. On the Transjordan plateau it obliterated many other Middle Bronze Age 2 cities, settlements, and nomadic tribes. This was a history changing event on many levels politically, culturally, and religiously as it destroyed kingdoms while allowing the flourishing of others.

# Chapter XI
## The 2013 Excavation Season

The eighth and last season of excavation at the site of *Biblical* Sodom called Tall el-Hammam before the publication of this book, took place from January 13 through February 21, 2013. The professional archeological staff was augmented by volunteers from the U.S., Australia, Great Britain, Canada and Germany as well as 30 local Jordanian workers.

The Kikkar stretches north to south from the throat of the Jordan valley to the beginning of the Dead Sea.

Tall el-Hammam is surrounded by smaller city and town ruins including Tall Tahouna to the northeast, Tall Barakat to the north, Tall Kafrayn located to the northwest, Tall Rama southwest, Tall Mwais south by southwest, and Tall Iktana south by southeast as well as other small towns. The ruin of Sodom is the connecting hub, positioned at the center of what was quite likely a confederation of city-states. These cities and towns looked to Sodom for leadership in a similar manner as the Greek city-states looked to Athens.

One thing all these Bronze Age sites have in common is the lack of any evidence of Late Bronze Age occupation. All occupation of these cities except Jericho, ended in Middle Bronze Age 2 circa 1600 + / - BCE. They were not reoccupied until near the end of Iron Age 1 and the beginning of Iron Age 2, after seven centuries of human abandonment.

The focus of the eighth season of excavation was done with a view of determining the relationship of Tall el-Hammam when it was still Sodom with the territory, cities, and towns under its cultural, military, and political control.

The excavation shows an evolution in Sodom's fortifications over time from Early Bronze Age 2 through Middle Bronze Age 2. Buildings and walls were in use for upwards of 2,000 years and not terminated until the mass destruction at the end of Middle Bronze Age 2. The oldest walls and dwellings are shown to extend both vertically and horizontally into the later walls and rooms.

The eighth season report also stated the city Early Bronze Age wall extended around the upper city, which was not thought to be the case until further excavation showed it to be fact. The report does not say the guard towers continued around the upper city, but that would seem to follow. I will show the upper wall defended by towers until I learn otherwise.[51]   It is evident by filled-in structures that the city had a number of gates later reduced to a single towered gate with a pillared gate house.  The gatehouse received a substantial amount of attention during this excavation season.

A group of circular stone-lined silos were found, indicating that grain was stored after the harvest for future consumption.  Early Bronze Age structures were found beneath some Middle Bronze Age walls with damage consistent with that caused by an earthquake(s) sometime in the past.

Middle Bronze Age ceramics are quite evident as are typical bronze tools and weaponry.

Intermediate Bronze Age pottery is abundant throughout the excavated site.  Structures inside the city walls are from the intermediate period, while buildings and walls from the Early Bronze Age were retained and reused by later inhabitants. The Early Bronze Age city is both substantial and quiet visible.  Pottery forms or the technique of making pottery from the earlier Kikkar city of Tulaylat al-Ghassul, believed to be the precursor of Sodom, is extant.

---

[51] One of the advantages to my books is their availability for corrections or additions because they are published on-demand.

The excavation report ends by recognizing that Sodom was one of the largest cities in extended Canaan from the Early Bronze Age until its destruction. That it was the central city-state in control of the Kikkar Valley and its travel routes. It also demonstrates the city was long lasting, existing for at least 1,500 years before its fiery demise. It further points out that its coalition with the other Cities of the Plain was unique in its region of the Levant.

# Chapter XII
## Future Excavations

### Models of Sodom

These are some of the discoveries I predict that will come from future excavations to the ruins of the Jordanic civilization in the Kikkar valley. It will be found or at least thought, that the actual name of Tuleilat al-Ghassul and the actual name of Sodom, are the same. The people took the name of their first city to their second.

I believe that relationships between the Cities of the Plain with Jericho and Salem were much closer than has been suspected to this date. It will be found that these two cities, like those of the Kikkar, were more deeply involved in the cultural, political, and religious aspects of Sodom than is believed today. Indeed, it may be discovered that many, and perhaps most, of the people of Sodom worshiped *Yahweh* the Storm God. Evidence, in the form of a small figure that the consort of *Yahweh*, Asherah, was a recognized deity in Sodom has already been found. And although one artifact does not a religion make, it points us in the right direction. Indeed, it is possible that the destruction of Sodom, attributed to *Yahweh*, may have actually provided the springboard that propelled his acceptance as the one almighty God. There is nothing like violent punishment by a god to focus the thoughts of those living near or within his base of devoted worshippers.

There is no doubt in my mind that Tall el-Hammam is the city called Sodom in *Genesis*. It follows that once a definitive date has been established for its destruction it will prove to be vastly important in the dating of all Ancient Near East events. History will change the view of major religions of today, to include Judaism, Christianity in all its forms, and Islam. It will be nothing short of a revolution in thought regarding the lives and cultures of ancient peoples.

As I mentioned earlier I think it will be demonstrated that the culture of the Kikkar was neither Canaanite nor Amorite in origin. It may even be shown to be the original native culture of the Neolithic and certainly the Chalcolithic peoples of the Levant.

The story unfolding at Tall el-Hammam will overturn the story of Sodom in *Genesis* and hopefully the view of Sodom and its allied cities in the future annals of history. I don't however expect it to change the way it is taught in many and perhaps most Christian churches and popular religious literature.

I believe it will be demonstrated that Jericho, beyond any doubt, ceased to exist as a viable city-state after the asteroid impact.

Future excavations I believe will also show that the cities and peoples of the Transjordan plateau perished at the same time as Sodom and the other Cities of the Plain.

With science and history tearing aside the shroud of time, the true story of Sodom, King Bera and Lot the Hebrew will be brought into the light.

## Models of the City of Sodom

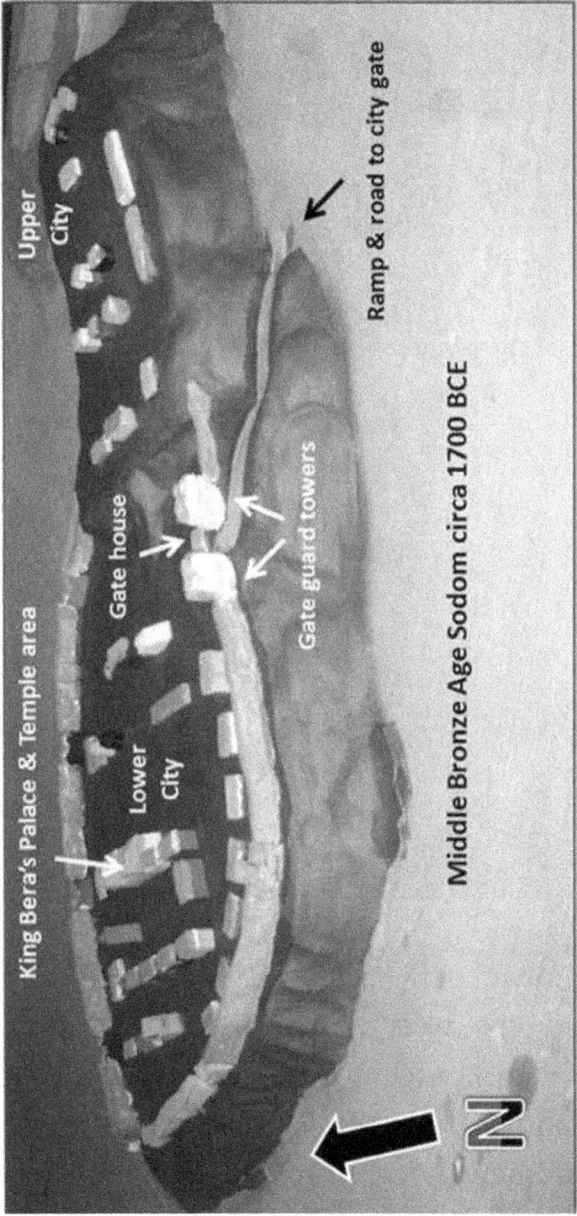

Upper City

Ramp & road to city gate

King Bera's Palace & Temple area

Gate house

Gate guard towers

Lower City

Middle Bronze Age Sodom circa 1700 BCE

N

Model of Sodom (above) – the overall city view before construction of the upper level wall.

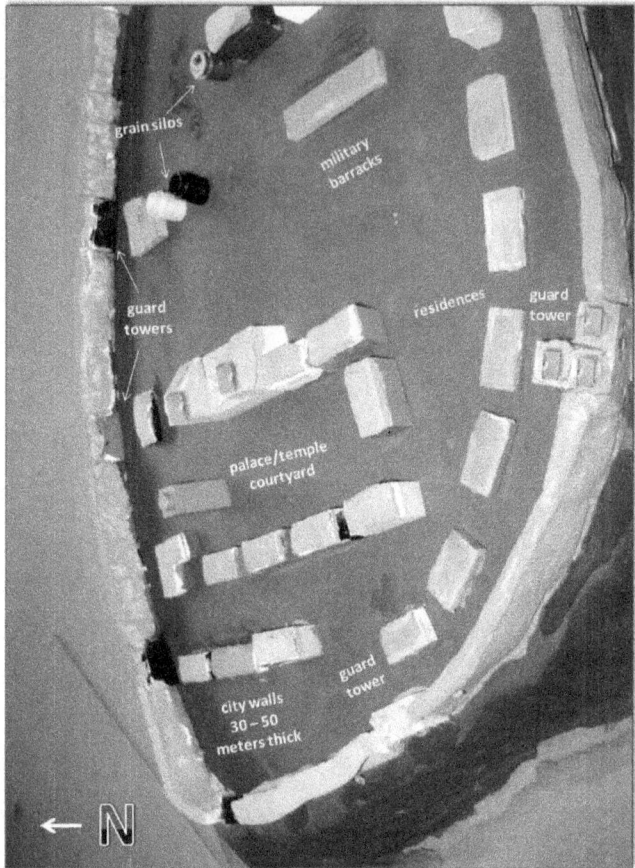

A model of Middle Bronze Age 2 Sodom showing the west end of the city proper, lower city. This model is based on drawings and descriptions from Tall el-Hammam excavation. The buildings were much denser in number than is shown above. The walls were an amazing 30 to 50

meters in thickness! City officials would have been housed in the lower city to include city elders (judges) such as the Hebrew Lot. The wealthier merchants would also have residences here. A military barracks provided the ever present City Guard and quick defenders for an attack from any quarter on the city walls. The soldiers also provided a labor force for agriculture when not on military duty.

A number of runners that acted as couriers when fast commendation was needed could keep contact on a daily basis with all of the Cities of the Plain. Jericho was the farthest away of the communities at about seven and a half miles from Sodom. Otherwise messages were taken during the normal course of the business day.

When a decision was necessary involving the various leaders, no doubt the Lord of the Kikkar, in this case King Bera would play host to his sub-kings in Sodom.

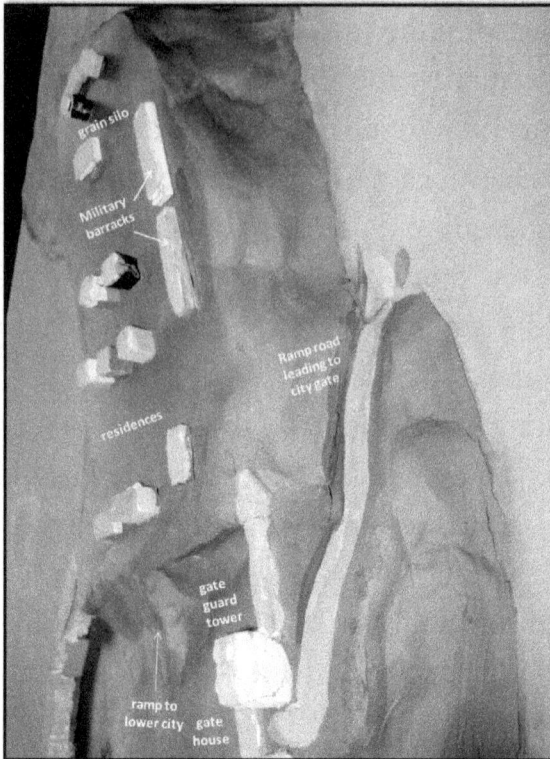

Model of Middle Bronze Age 2 Sodom showing the east end of the city proper, upper city.

The buildings were much denser in number than is shown above. The more common citizens of Sodom were housed in the upper city to include servants, the farmers that worked the inside city agriculture fields, lesser merchants, and tradesmen. Lot's observation post was near the gate and likely in the gatehouse itself where he could watch the comings and goings of city visitors. This model was constructed before the latest field reports stated that the upper city did indeed have a wall around it.

There are two military barracks in the upper city to afford quick defenders to the towered city gate. Somewhat higher and therefore harder to attack, the city walls did extend to the upper city, making it even more secure.

# Chapter XIII
## After the Death of Sodom

In his book Dr. Collins stated that from a biblical point of view the death of Sodom is quite clear, deliberate and divine retribution.

After weighing the evidence accusing evil-doing leveled against Sodom and Gomorrah and indeed all the Cities of the Plain, the hamlets, the villages and the individual farmsteads, I have a different point of view. I find the evidence for evil on the part of these people greatly lacking or over inflated. I also find the evidence appointing *Yahweh* as the killer of all these Canaanite and Amorite people lacking. In other words I don't think the deity of Israel did it at all.

I have already stated that Abram departed precipitously for the southern margins of the Land of Canaan according to the account in *Genesis*. But I also feel that before he left Abram consulted with King Melchizedek the high priest of *Yahweh* regarding what happened in the Kikkar Valley. For it seems to me that any of us, faced with such senseless and total devastation would seek answers from a spiritual leader.

Without any doubt Melchizedek seized the opportunity to solidify the belief in *Yahweh* as the one and only almighty deity in the mind of Abram who in actuality was a new convert in his belief of the Hyksos Storm God.

It seems apparent to me that the destruction of the Jordanic Civilization launched, or at least gave great impetus, to the belief in *Yahweh* and the bulk of the world's belief in a one, almighty deity.

Thankfully the Hebrew story tellers, religious men and eventually the writers of the *Bible* kept the story of Sodom alive, albeit negatively, for millennium. Otherwise the memory and narrative of the only known city on Earth destroyed by an asteroid would still be unknown.

Are there possibly others whose memory has been lost, changed, or ignored by modern readers? I have a feeling there are. So, what are the chances an event that mirrors the death of Sodom could happen today? I will address that in the second part of this book.

What is glaringly obvious is that with the elimination of Sodom, King Bera and his fledgling hegemony in the Kikkar, there would never be a Canaanite empire. The fertile lands of the Jordan River, unlike those of the Euphrates and Nile, were not allowed to give birth to the future it once promised. Not until King David united the Hebrew tribes seven hundred years later[52], would anything like a true nation-state rise in the Land of Canaan.

There can be no doubt that on a warm summer morning nearly four thousand years ago, Abram stood hard by Hebron, looked across the expanse of the Dead Sea and down into the verdant valley of the Jordan River. There, his eyes, following a fiery trail marked by roiling black and grey smoke and lit up with the light of a dozen suns, watched the death of not just cities and a budding empire, but the very birth of a single almighty deity!

To clearly answer the question posed in the title of this book *Did God kill the king of Sodom?* I feel compelled to answer with an emphatic *no*!

The final chapter in this book will, I believe, present you, the reader, with the smoking gun in this mystery. Or perhaps I should say the smoking asteroid.

---

[52] Approximately 1000 BCE *Archeology in the Holy Land* p. 102.

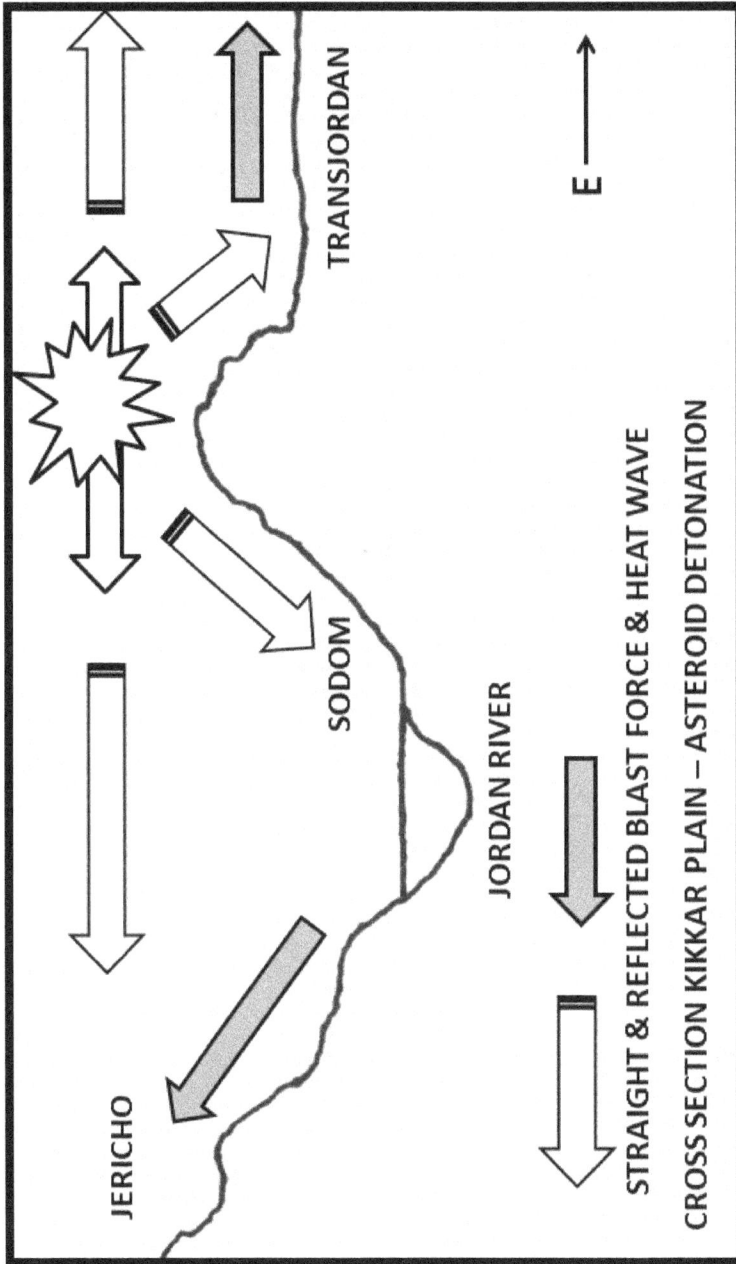

CROSS SECTION KIKKAR PLAIN – ASTEROID DETONATION

STRAIGHT & REFLECTED BLAST FORCE & HEAT WAVE

JORDAN RIVER

SODOM

TRANSJORDAN

JERICHO

E

# Part II
# Lessons for the 21st Century

# Chapter XIV
## Some Known Asteroid Air Bursts

Over the past century many explosions in the air caused by the entry of asteroids into the atmosphere of Earth have been recorded. By far the most devastating and famous is known as the Tunguska Event. Luckily, that air burst took place in a relatively remote area from human cities and towns, but others, admittedly of a lesser magnitude, have happened closer to population centers.

The information gleaned from these events will provide a better look and understanding of what happened to the Cities of the Plain nearly four thousand years ago.

### Fireball over Spain

On the night of December 8, 1932 under a clouded sky with rain, a large fireball bright enough to be seen through the clouds streaked across the sky above a village in southwestern Spain. Many people in the remote town of Arroyo-molinos de Leon witnessed the event. The asteroid approached the Earth almost straight down at an estimated 80 degree angle. It exploded in the air with a deafening roar and the concussion caused considerable physical damage to the buildings and residences of the town. There was no crater or fragments of the meteor afterward indicating it was completely destroyed in the airborne explosion.

The velocity was estimated at 14 kilometers per second or 31,317 miles per hour! The size of the meteor is estimated to be 18 meters or 19.7 yards wide. Luckily for the people of the village the meteor exploded ten miles up, because the energy released by the explosion developed the equivalent of 190 kilotons of TNT. Obviously, had the air burst occurred closer to the ground the damage and loss of life would have been catastrophic for the small village and its people.

When you consider that Nagasaki, Japan was destroyed by an atomic bomb that generated an explosive force of 21 kilotons of TNT, the importance of the height of the explosion becomes clear. The Nagasaki bomb was detonated at the low altitude of 1,650 feet, while the meteor over the village exploded at 52,800 feet. That is more than 32 times higher than the atomic bomb. Without a doubt Arroyomolinos escaped destruction by just a matter of seconds. But the two most spectacular modern day meteor events both took place in Russia, and one of them happened in quite modern times, in fact, in the 21$^{st}$ century.

## The Tunguska Event

In the early morning hours of June 30, 1908 an asteroid streaked through the sky above the Tunguska River and forest in Siberia. It came in at such a shallow angle that it got close to the Earth very quickly, so close that witnesses to the event could feel a blast of heat as it passed by them overhead.

The asteroid detonated above the forest, blasting it with searing heat that set the trees ablaze, followed instantly by a concussive shockwave that temporarily blew out the fire and knocked literally millions of trees to the ground in all directions from ground zero. Then the interrupted fires returned.

U.S. researchers have estimated the detonation of this meteor approached the 5 megaton (5 million tons of TNT). Such a detonation is equivalent to 1,000 atomic bombs! The airburst flashed down in a fireball that exceeded 25,000 Kelvin, that is equivalent to more than 44,000 degrees

Fahrenheit! Herds of deer and elk were scorched to death or vaporized while trees were incinerated or ignited out to a distance of more than 800 miles. In the words of one researcher it was the loudest, fiercest, most forceful explosion ever known to civilized man.[53]

After the event cities as far away as London experienced bright nights for many weeks. The night sky was lit up so well that photographs could be taken at midnight as the light continued around the clock. What causes this phenomenon is unknown but scientists speculate it was dust particles, water vapor, or perhaps ice crystals in the air excited on the molecular level by the event.

The asteroid, thought to have been made up primarily of stone, came into view of witnesses as a blinding object traveling at an unearthly speed with a five hundred mile trail of smoke behind it. The explosion was heard hundreds of miles away in the capitols of Europe. The recent drilling of tree cores from trees that survived the event has yielded fourteen isotopes known to exist in stony asteroids.

It was 19 years before any real investigation was mounted by the Russians to try and find out what had happened during the event. Even almost

---

[53] The exception of course, is the Sodom event of 1700 BCE.

twenty years after it happened the destruction was quite evident with trees downed for a radius of hundreds of miles.

When the expedition approached the local native people to guide them to the site they were reluctant. The reason given was that they feared the cosmic event was visited upon them by their fire god named Ogdi, and if they approached they would be punished. This is very similar to the way the Hebrews and other Canaanite peoples viewed *Yahweh* after the Cities of the Plain were destroyed. For centuries afterward the area was viewed as a dangerous and sacred place that was visited only for certain religious ceremonies. The Egyptians referred to the remains of Sodom as the place of "great mourning".

At Tunguska the locals recounted that during the event their tents were blown away by a fierce hot wind, their clothing was ignited, and fires were lit spontaneously around them. They believed at the time that the end of the world had come.

Researchers believe that the asteroid was about 100 feet wide, but that the same kind and magnitude of event could be triggered by an asteroid as small as 20 feet across. That dramatically widens the number of possible future catastrophic asteroid impacts on Earth.

Researchers estimate that an asteroid similar in size to the Tunguska asteroid enters Earth's atmosphere every hundred years. It has been more than a century since Tunguska, and an event approaching its magnitude is overdue.

## Super fireball over Chelyabinsk

February 15, 2013. The official name of this meteorite is Chelyabinsk 2013. The Chelyabinsk Event is one of the largest recorded meteor air bursts in modern times, and thanks to car dash cameras and hand held camera phones the first seen by millions of people. And yet the Lunar and Planetary Institute determined this meteorite is on the extreme small end of this type of event. The typical impact velocities of chondrite asteroids entering the Earth's atmosphere is 40,000 plus miles per hour! Speed plus low altitude burst level equals massive and immediate destruction.

If there were any skeptics about asteroid explosions on or near earth, they were convinced after viewing this event. The asteroid passed near the Russian city of more than a million persons traveling at the astounding speed of more than 41,000 miles per hour. The light from the fireball appeared brighter than the rising sun and caused objects on the ground to cast shadows. Although passing over the city at least fourteen miles above, people on the ground felt an intense heat wave given off by the flaming asteroid as it passed.

The asteroid came in at a shallow angle of entry and provided the world with a brilliant display of fiery, awe-inspiring power. It exploded very high at fourteen and a half miles (76,000 feet) above the cities on the plain. Had the angle of descent been more vertical, the destruction that happened would have increased exponentially.

Despite this extremely high altitude 1,500 people were injured to the extent that they required medical attention. All recorded injuries were from flying debris created by the powerful shockwave. Since the explosion occurred very high up, the atmosphere had time to absorb the greater portion of the 440 kilotons of TNT blast[54] before it reached the ground. None-the-less more than 7,000 buildings in Chelyabinsk and four other cities[55] [sound familiar?] were damaged by the massive shockwave. Not all the energy was released in the blast, with 90 kilotons of TNT used to generate visible light.

---

[54] 21 times the amount of energy produced by the Fatman atomic bomb.
[55] Orenburg, Bashkortostan, Sverdiovsk, Tyumen.

Thousands of asteroid fragments showered around the villages of Pervomaiskoe, Deputatsky, and Yemanzhelinka 40 kilometers south of Chelyabinsk.

Although this meteor was about the same size as the Arroyomolinos meteor (17-20 meters), it appears to have been made up of sterner material. Recovered pieces of the object allowed its identification as an L Chondrite meteorite mostly composed of hard stone with 10% of iron mixed into it. Its estimated overall mass was determined by NASA scientists to be 11,000 tons, much larger than the estimated 2,500 tons of the Arroyo-molinos meteor.

The asteroid came from out of the sun and so was hidden from view until it burst upon the plain around Chelyabinsk. Route of travel calculations for the asteroid indicates that it came from a cluster of NEAs (Near Earth Asteroids) named for the Greek god Apollo. The orbit of these NEAs is elliptical and takes them into the path of Earth's orbit.

Prior to the Chelyabinsk Event, the largest chondrite asteroid detonation was over the Gold Basin Area[56] in Arizona. Only eight meters in size

---

[56] Located in the north west corner of the state.

this meteorite developed between 5 and 50 kilotons of TNT explosion.

# Chapter XIV
## Earth's Detection Programs

The United States *Satellite Early Warning System* is comprised of a number of reconnaissance space craft under operational control of the U.S. Air Force.

Don't feel too secure yet, because these satellites are part of the *Defense Support Program* and are not exclusively dedicated to the surveillance of Near Earth Objects (NEOs). They have the additional mission of watching the Earth's surface to detect and track the launch of rockets or missiles. Placed in stationary geosynchronous[57] orbit, they are equipped with infrared sensors inside a wide angle *Schmidt* camera, also sometimes referred to as a *Schmidt* telescope. This telescope is based on technology developed in 1930 by Bernhardt Schmidt.

---

[57] An orbit around the Earth that matches the Earth's own rate of rotation.

Photo: Mogens Engelund.   Photo of a *Schmidt* telescope-camera in use in 1966 at the Brofelde Observatory in Denmark.

The 21st Space Wing is stationed at Peterson Air Force Base in Colorado operates satellites of the DSP.   Warnings of missile launch or the observation of Near Earth Objects are forwarded to NORAD, (North American Aerospace Command) in El Paso, Texas or to USSTRATCOM; (United States Strategic Command) at Cheyenne Mountain, Colorado.

The data received is immediately reported to appropriate agencies and commands around the world for information and possible response.

USSTRATCOM

The number of DSP satellites is classified, but over the program's lifespan they have launched twenty three satellites, however the primary mission remains: *Strategic and tactical missile launch detection.* And is not the observation of Near Earth Objects, those naturally occurring asteroids that can become missiles of mass destruction without any warning at all.

An ongoing program is SIBRS-High or Space Based Infrared Systems is under development, but has been hit with high cost overruns, technical problems and doubts about its operational capabilities. All of this, at a time that Earth seems to be coming under a seemingly growing threat originating from outer space.

NASA NEO Program

Spaceguard NASA's Near Earth Object Program

Through their NEO Program dubbed "Spaceguard" NASA detects, characterizes, and tracks asteroids and comets using both space and ground based telescopes, to include those with an infrared capability. The Jet Propulsion Laboratory (JPL) located in California manages the NEO Program for NASA.

As recently as October, 2013 a newly discovered asteroid dubbed 2013 TV135 was detected by Ukrainian scientists at the Crimean Astrophysical Observatory and reported to NASA and the rest of the world.

The asteroid made a surprise close approach to Earth on September 16. Estimated at 400 meters wide, this asteroid is *twenty times* larger than the one that struck the Chelyabinsk area in Russia. Had it entered the atmosphere it too would likely have exceeded a velocity of more than 40,000 miles per hour and caused damage far greater than its smaller companion at Chelyabinsk (20 meters wide).

Asteroid 2013 TV135 is on a rendezvous with Earth's orbital neighborhood in 2032. NASA states that well before that date the precise orbit of this asteroid will be known and the likelihood of its striking Earth is less than 1%. I guess we'll find out for sure in 2032, so you have nineteen years to build yourself a bunker if you are a doomsday prepper[58].

But, if it did become a PHA (Potentially Hazardous Asteroid) would we be able to defend ourselves or our planet from it?

In March of 2013 the House Committee on Science, Space, and Technology held hearings on a plan of action against what it termed *space threats* especially from asteroids. The hearings were in direct response to the Chelyabinsk event the month

[58] Also known as survivalists, "doomsday preppers" is considered by some as somewhat of a derogatory term.

before and the near approach of a second asteroid that same day. Asteroid 2012 DA14[59] came inside the orbit of the moon and approached to within 17,100 miles of the Earth. That number may sound large, but in cosmic terms, that is a *near miss*. This asteroid was more than twice as large as the Chelyabinsk asteroid. Had it struck Earth, it would have generated an explosion of nearly 2.4 *megatons* of TNT.

The committee solicited comments from the White House, NASA, and the Air Force.

U.S. House Representative Lamar Smith of Texas began by observing that an asteroid as small as 100 meters in size could destroy an entire city with a direct hit. But actually an asteroid could be one fifth that size and do the same thing depending upon its angle of descent, velocity, and the height of its detonation.

Using an online academic webpage called "Nukemap" I detonated an explosion of 50 kilotons, equal to the smallest recorded chondrite asteroid impact of 17 - 20 meters[60], over a densely populated major Earth city. The effect on the population alone was the immediate death of

---

[59] This designation indicates that this asteroid was only detected in 2012.
[60] A reference to the Gold Basin impact in Arizona 50,000 years ago.

226,000 people with 350,000 people injured bad enough to require emergency medical care.

In response to the chairman's question John Holdren, the White House senior advisor for science and technology, said that scientists could not stop an asteroid from striking the earth and that the number of potential city killing asteroids was very large.

NASA administrator Charles Bolden simply recommended that the space agency continue to do research on the problem.

General William Shelton, U.S. Air Force, admitted that Air Force Space Command wasn't actually paying much attention to NEOs other than manmade objects in orbit. His words to the committee were:

*...Air Force Space Command sensors were developed to track manmade objects in Earth orbit. The Nation's current capability to track asteroids, which orbit the sun, is largely driven by NASA.* [This is a reference to NASA's Near Earth Object Program Office at the Jet Propulsion Laboratory].

The general then went on to discuss the future replacement of the DSP satellites by another system he called JMS as in JSpOC (Joint Space Operations Center) Mission System. He made no mention of SIBRS-High which may indicate the termination of that program. He said the Air Force Space Command has forces and capabilities to detect, track, identify and characterize manmade objects in Earth orbit.

He said nothing about any capability to engage, destroy, or redirect any space object or objects. What wasn't said, says volumes about the lack of any real defense for the Earth. It seems to me, we need leaders with a wider view than those in charge of our space command today seem to have to offer. Of course as commander in chief, the president would have to step forward and add this mission to our military.

European NEOShield Program

The following statement is from the homepage of the NEO Shield Project's website.

*The NEOShield project has been set up to carry out a detailed analysis of the open questions relating to realistic options for preventing the collision of a NEO with the Earth. The aim of the project is to provide solutions to the critical scientific and technical issues that currently stand in the way of demonstrating the feasibility of the promising mitigation options with a test mission. While a mitigation test mission is beyond the financial scope of the current project, the NEOShield team aim to provide detailed test-mission designs for the most feasible mitigation concepts, so that it will be possible to quickly develop an actual test mission at a later stage.*

*The project concept includes laboratory experiments and associated modelling [sic] of the behaviour [sic] of a NEO during a deflection attempt. These experimental results and modelling [sic] will help to improve our understanding of the nature of NEOs and allow the feasibility of mitigation techniques and mission designs to be accurately assessed.*

*As well as the European team, the project includes partners from established space-faring nations outside the European Union to allow the team to formulate a global response campaign roadmap that may be implemented when a serious impact threat arises. The roadmap will consider the necessary international decision making milestones, required reconnaissance observations, both from the ground and from rendezvous spacecraft, practical prerequisites, such as precise orbit tracking, and a campaign of perhaps several mitigation missions, depending on circumstances.*

The Canadian Space Program

Headquarters for the Canadian Space Agency is located in the John H. Chapman Space Centre (sic) in Saint Hubert, Quebec. The CSA was established and funded on May 5, 1990.

The CSA joined the search for asteroids in February of 2013 with the launch of the Near Earth Object Surveillance Satellite (NEOSSat). NEOSSat is a cost effective microsatellite[61] using a Maksutov (named after its Russian inventor) design telescope to search for interior-to-Earth-orbit (IEO) asteroids.

---

[61] Any satellite under 500 kilograms of weight is classified as a microsatellite.

Dubbed NESS for Near Earth Space Surveillance, the mission is to search for and track asteroids in orbit primarily between the Earth and the Sun. Dr. Alan Hildebrand of the University of Calgary is the mission director.

These asteroids come primarily from the Aten asteroid group. Although these asteroids are usually circling between Earth and the Sun their furthest swing outward brings them temporarily into Earth's path. At that time they can pose a threat of impact. Here is a partial list of the more than 730 asteroids in this group, many of them as large as or larger than the Chelyabinsk asteroid which was only 20 meters in diameter.

The largest and ironically the dimmest is Hathor 2340 with a diameter of 5.3 kilometers. The brightest is unnamed but with the designation of 1999 HF1 with a diameter of 4.3 kilometers. Then comes 1998 TU3 at 3.6 km, 3753 Cruithne 3.3 km, 3362 Khufu at 700 meters, 2001 KB67 505 meters and SW130 at 5 meters, the only one in this list smaller than the Chelyabinsk asteroid. This means that 728 of the interior asteroids are larger than the asteroid that struck over Chelyabinsk.

As this partial list of asteroids demonstrates, this is an important mission crucial to the detection of asteroid threats to Earth.

The second group of asteroids NESS looks for come from the Atira or Apohele[62] asteroids that orbit completely inside Earth's orbit and pose a less but not impossible threat of impact. These asteroids can be perturbed by the gravity pull of Venus and Mercury, flung outward and possibly into the path of Earth. There are only twelve known Atira asteroids.

Many of these asteroids are hard to detect as they rocket toward Earth out of the glare of the Sun. Although a low cost asteroid surveillance system, NESS is an important contribution to the asteroid surveillance program.

---

[62] Apohele is Hawaiian for "orbit".

I consider the B612 Foundation a very important future contributor to Earth's defenses against a surprise asteroid impact through their space-borne mission called Sentinel.

The B612 webpage describes their project as follows:

*Sentinel is a space-based infrared (IR) survey mission to discover and catalog 90 percent of the asteroids larger than 140 meters in Earth's region of the solar system. The mission should also discover a significant number of smaller asteroids down to a diameter of 30 meters. Sentinel will be launched into a Venus-like orbit around the sun, which significantly improves the efficiency of asteroid discovery during its 6.5 year mission.*

The B612 space craft and instruments use NASA proven deep space systems that will be able to look at the Sun to detect Near Earth Objects. The systems include large space-based telescopes of both the Kepler and Spitzer designs and a large format camera. After launch in 2018 it is expected to detect and track nearly all of the NEOs greater than 50 meters in diameter.

The data gathered by Sentinel will be used to create a map of the inner Solar System. The map will allow both robotic and manned mission exploration of this vital area of space. Sentinel will also identify potentially dangerous space objects and provide an early warning of impending asteroid impacts coming from the direction of the Sun. The foundation made the following statement regarding Sentinel:

*The B612 Sentinel mission extends the emerging commercial spaceflight industry into deep space – a first that will pave the way for many other ventures. Mapping the presence of 1,000's of near earth objects will create a new scientific database and greatly enhance our stewardship the planet.*

The foundation asserts the Sentinel map will give us enough advance notice of an impending impact that a deflection mission can be launched.

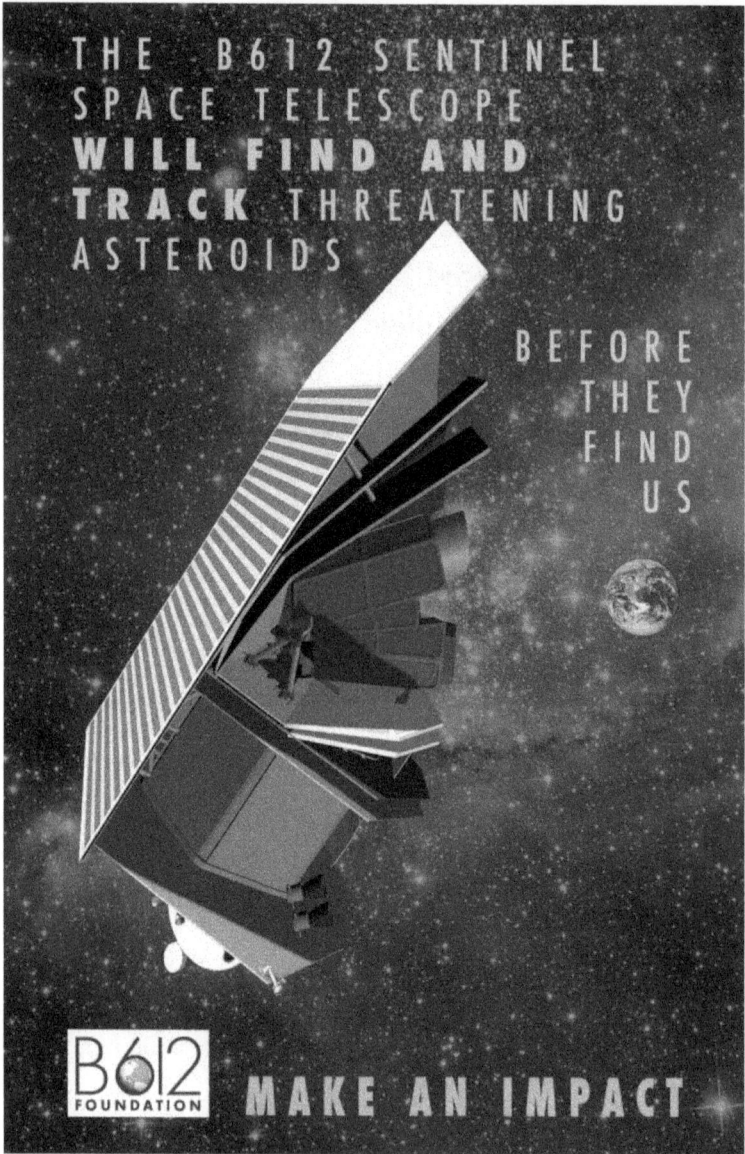

The B612 Foundation's Sentinel poster.

The B612 Foundation is a publicly funded project. If you go to their webpage you will find directions on how to make a donation or how to purchase one of their store products. My favorite is a coffee mug that reads "Don't be a dinosaur" with an asteroid in place of the letter "O". Beneath the message is a dinosaur peering through a telescope and saying "All clear guys."

Here is the site's URL: https://b612foundation.org/

**Israel Space Agency**
**סוכנות החלל הישראלית**

# Israel's Space Agency

The Israel Space Agency was established in 1983 and has launch capabilities. Although it has not launched any known NEO satellites, it does operate the National Knowledge Center on Near Earth Objects.

The ISA formed and operated the center at Tel-Aviv University to study minor bodies to include asteroids in the solar system. Its purpose is to map any object that may pose a threat to Earth and to find a way to eliminate the threat.

Without any satellites the program operates two telescopes at the Wise Observatory. To advance the program a special wide-field 46 centimeter telescope was obtained and operates at the observatory. To date the telescope has been successfully used to locate several tens of newly discovered asteroids. The research focus of this program is on the rotational properties of NEOs with a concentration on asteroids through the investigation of their light curves. The funding from the ISA has been terminated but observations continue unabated.

Asteroid light curves are the graphing of changes over time of asteroids to determine their shape and axis of spin (rotational period). This graphing is done to support radar observation or analysis of the object's threat to Earth.

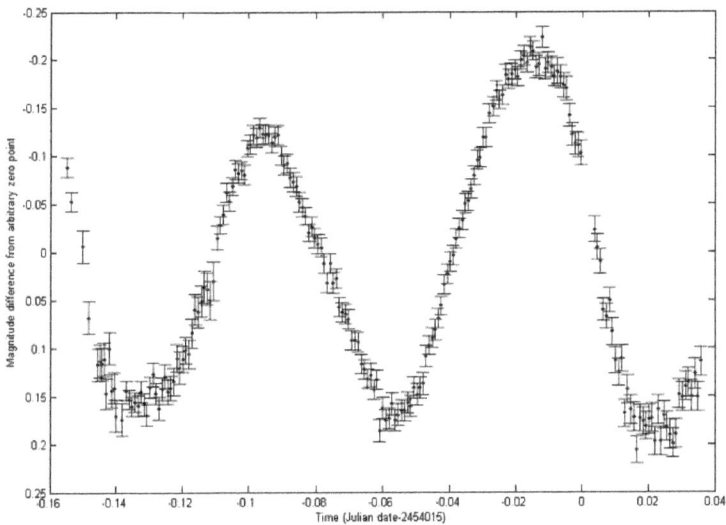

Light curve of the asteroid 201 Penelope based on images taken on 6 October 2006 at Mount John University Observatory. Courtesy Mount John University.

## Earth's Defenses

Although a number of ideas have been offered, there is no concerted plan by Earth's governments to defend ourselves from a planet killing asteroid. In that regard, despite our greatest technological advances, we are as vulnerable to extinction as were the dinosaurs of 65 million years ago!

Not even the United States' touted Aerospace Force, better known as the U.S. Air Force, has a plan of defense. Unfortunately, our seemingly weak national government under the leadership of President Obama is too wrapped up with other concerns to even consider funding a defense. In the aftermath of the Chelyabinsk near disaster there was a flurry of activity that amounted to a lot of political smoke and steam with no real fire or zest for this quite obvious and possibly growing problem.

It will take a disaster before the threat is taken seriously or anything will really be done in the area of a planetary defense. But all we have to do is look back at the fate of Sodom, Gomorrah, Jericho and the other cities of the Jordan Valley to understand what can and has happened.

Do I have a recommendation? Yes, in fact I do. If you have a science oriented son or daughter with a birthday coming soon, or other gift giving holiday, I suggest you get them a telescope or set of binoculars, preferably with an infrared capability. We need more eyes on the skies than ever before.

# About the Author

Walt Cross holds degrees in arts, science, and history from Oklahoma State University and the University of the State of New York. He is also a graduate of the United States Army Sergeants Major Academy. He is known as the "unruly historian" because he does not concentrate his writing inside one particular genre but writes about what interests him. Some of his other titles include:

*Custer's Lost Officer; the Search for Lieutenant Henry Moore Harrington, 7th U.S. Cavalry*

*From the Beaches to the Baltic; the Story of the 7th Armored Division in WWII* [contributing editor]

And many other history titles available online at www.lulu.com/greenpheon7.

# Name Index[63]

---

[63] For individuals known by more than one name I have placed the most accepted name first.

# Bibliography

COLLINS, STEVEN and SCOTT, LATAYNE 2013. *Discovering the City of Sodom.* Howard Books, a division of Simon & Shuster, Inc. New York, NY

GLUECK, NELSON 1940 *The Other Side of the Jordan.* American Schools of Oriental Research, New Haven, Connecticut

KENYON, KATHLEEN 1960 *Archaeology in the Holy Land* Frederick A. Praeger New York, NY

KENYON, KATHLEEN 1966 *Amorites and Canaanites* Oxford University Press London, England

NEEV, DAVID and EMERY, K. O. 1995 *The Destruction of Sodom, Gomorrah, and Jericho* Oxford University Press London, England

www.ingramcontent.com/pod-product-compliance
Lightning Source LLC
Chambersburg PA
CBHW031252090426
42742CB00007B/415